IN THEIR RIGHT MINDS

The Lives and Shared Practices of Poetic Geniuses

Carole Brooks Platt

imprint-academic.com

Published in the UK by
Imprint Academic, PO Box 200, Exeter EX5 5YX, UK

Distributed in the USA by
Ingram Book Company,
One Ingram Blvd., La Vergne, TN 37086, USA

ISBN 9781845407896

A CIP catalogue record for this book is available from the
British Library and US Library of Congress

Cover artwork by Jay Vogelsong

For Mother

Acknowledgments

I have greatly benefitted from the knowledge of professors, colleagues, friends and family who have shared their knowledge, and pointed me in directions I might not have discovered on my own. Many thanks to Michèle Sarde and Patrick Brady, who first introduced me to literary critical methods, and my sister, Janice, for the reference to Robert Graves, whose matriarchal mythopoeic theme became the underpinning of my doctoral dissertation, and recurred again in this study of poets.

I first became aware of hemispheric differences that could explain the voices and visions of poets and prophets in Julian Jaynes' *Origin of Consciousness in the Breakdown of the Bicameral Mind*. Marcel Kuijsten, director of the Julian Jaynes Society, suggested I attend the Toward a Science of Consciousness conference and present my research there. This venue was an invaluable entrée into a world of minds intent on discovering the underpinnings of consciousness, the effects of altered states, and the possibilities of precognition and non-local awareness. Special thanks to Dr Allan Schore, who introduced me to the mother's import-ance in establishing a secure sense of self, and provided me with a steady stream of references and articles confirming the effect of maternal attachment on the early developing brain.

My chapters weave together, amplify and update materials I have already published in *Gnosis, Journal of Consciousness Studies, Jaynesian, Clio's Psyche* and *Plath Profiles*, to whom I am grateful for publishing my research. Thanks as well to the many theorists whose lectures and conferences at the Jung Centers of Houston and New York who have inspired me and taught me so much.

Specialists who read my drafts and/or offered invaluable suggestions for further readings or factual changes include Julie Kane, James Hollis, Judith Moffett, Timothy Materer, Dianne Hunter, Jeffrey Kripal, Wouter Hanegraaff and Lois Oppenheim. Many thanks to the hundreds of neuroscientists whose studies have helped clarify the functional differences between the left and right hemispheres and the effects of atypical lateralization on the brain.

Thanks to my dear friend whose personal experience jettisoned me on a path to understanding how atypical minds and traumatic circumstances can change the way the brain works, with enormous creative potential.

Finally, I am deeply indebted to my husband, Charles, and my sons, Colin and Jonathan, who have supported me throughout this twenty-year journey, pushing me to complete the work and cross the finish line.

Contents

Introduction   1

Chapter 1. Art, Music and Poetry in Atypical Minds   3

Chapter 2. Sensing Presences   24

Chapter 3. Emotion, Dissociation and Linguistic Creativity   40

Chapter 4. The Medium and the Matrix   60

Chapter 5. From Myth to Mediumship   81

Chapter 6. Dictating Others and Surrogate Mothers   106

Chapter 7. A 20-Year Ouija Board Odyssey   132

Chapter 8. Right Hemisphere Imbalance versus Espousal   153

Conclusion   179

Bibliography   183

Introduction

My interest in the voices and visions of poets and prophets was precipitated by a dear friend's claim to channel angels after her mother died. In the face of a real human being, whose friendship I treasured, not an anonymous participant in a scientific study, how could I not relentlessly pursue the reason for her puzzling transformation?

The quest began with scientific and mystical readings, but soon branched out into a study of the lives of great poets, whose creative productions flowed from techniques like automatic writing, séances or the Ouija board. Most of them required collaborating partners to help contact 'discarnate' entities who 'spoke'. How could these intelligent people believe what they were doing was real? How did it work?

In 1976, Julian Jaynes proposed that the language of poetry, with its different tenor and tone and use of unusual metaphors, as well as prophecy, with claims predicting the future, originated in the right side of the brain. Poetry was the language of the gods. Neuroscientific evidence from the twenty-first century, especially Julie Kane's use of organic brain studies along with her own knowledge as a scholar/poet herself, proved that poetry is right-hemispheric language. Studies on individual aspects of poetry, such as use of vowels versus consonants, rhyming, willed or unwilled production, show how hemispheric differences do exist. Psychological investigations connecting mood disorders and creativity still prevail, but rarely make the enhanced right-hemispheric connection.

It is well known that left-hemispheric dominance for language is the norm. Hoping to avoid distortion in their results, most fMRI studies on the brain only include right-handers, neglecting left- or mixed-handers. But other researchers in lateralization studies have found that *reduced* cerebral asymmetry, with right, bilateral or even synchronous activation of the hemispheres, can produce not only poetry, but also a sense of presence *and* belief in the paranormal, with or without mental imbalance. An atypically enhanced right hemisphere, then, whether attained through genetic predisposition, left-hemispheric damage, epilepsy, autism, childhood or later traumas can create hypersensitivities along with special skills. Dissociative 'others' may arise unbidden or be coaxed out through occult practices.

The poets included in this study are Blake, Keats, Hugo, Rilke, Yeats, Merrill, Plath and Hughes. Blake took 'divine' dictation, as did Rilke at times. Keats, the 'chameleon' poet, merged, and is included in this study because of a dream invitation that responded to my concerns for my friend. All of the others had collaborating partners, one of whom was the transcriber, while both were needed to allow the entities to 'speak'. Based on nearly twenty years of scientific and literary research, this book enters the atypical minds of poetic geniuses — gauged by visible signs and invisible indications in their lives — attempting to make sense of poetic creativity and paranormal claims.

What could still be perceived as direct dictation in the eighteenth century, by the nineteenth required techniques and collaboration with supportive others. My analysis makes clear that personal as well as historical tragedies could constellate to produce a psychological environment conducive to these dissociative experiences, while providing answers to deep existential issues.

The mythopoeic imagination seems to lie in unconscious processes, ready to come to the fore when the stresses of life become overwhelming. Are the voices real? Whether heard or spelled out by dissociative means, they are sculpted from personal and collective raw materials, speaking to the needs of the moment.

This subject is vast, but really encompasses one basic phenomenon: the astonishing ability of the human mind to shape its chaotic electrochemical underpinnings into creative solutions for survival. The teeming, energetic, unconscious undergrowth can be harnessed, bringing forth form which can guide, instruct, warn and give meaning to battered souls or to whole communities and nations.

Not everyone is capable of hearing the voices and giving them expression. Poets, artists, mystics and madmen, whose minds are organized in a different way from others, can break through the barrier of everyday perception to get glimpses of immaterial truth or beauty and give it form. As a Tibetan Buddhist monk once said to me when I asked him about my friend's experience: 'There is a place where created and real meet.' I found that answer consoling, but the rest relied on diligent research and a sympathetic heart.

Chapter One

Art, Music and Poetry in Atypical Minds

Then first I saw him in the Zenith as a falling star,
Descending perpendicular, swift as the swallow or swift;
And on my left foot falling on the tarsus, entered there.
— William Blake

Inner Voices

Most people hear an inner voice, recognized as their own, on a daily basis. A positive voice can motivate us, narrate the day, suggest future strategies as well as recall relevant past experiences for comparison purposes. Sometimes we speak our thoughts out loud when alone to encourage us or calm our emotions. This inner speech is a large part of our conscious awareness and helps maintain our sense of self. But what part of the brain is speaking to us? Ferris Jabr (2014) reports that Broca's area in the frontal lobe, responsible for producing speech, sends commands backward to Wernicke's area, on the border of the temporal and parietal lobes, necessary for understanding speech. Receiving the message, Wernicke's area does not respond as though it were someone else's voice. But, a weaker than usual electrical signal traveling between the front and back language areas will be experienced as auditory hallucinations.

In *The Origin of Consciousness in the Breakdown of the Bicameral Mind*, Princeton psychologist Julian Jaynes (1976/1990) used a right-to-left hypothesis, claiming that prior to 1000 BCE the human race could only react to their environment. Not yet fully conscious as modern humans, they lacked the ability to introspect their thoughts. Hallucinatory voices guided their actions when the stress of novel situations required a creative solution. Jaynes said the 'alien' voices were projected to Wernicke's area on the left ('man' side) of the brain from the corresponding area on the right ('god' side). The god-like speech mimicked the authoritarian voices of parents, tribal leaders and kings. Professor and poet Julie Kane (2004) wrote that preliterate peoples, illiterates and young children evidence right-hemispheric processing of language, which switches to left-hemispheric dominance after acquiring written language skills. It makes sense the right hemisphere, more important for navigating their hostile environment, was more dominant.

While the late date of conscious origins is certainly debatable, Jaynes's brain lateralization theory, parts of which have been confirmed by more recent neuroscientific research, makes it pertinent to this present study of nineteenth- and twentieth-century poets who consciously or unconsciously used dissociative means to access words from inspirational 'Others', with knowledge seemingly beyond their own. Jaynes believed that certain modern era poets retained residues of the so-called 'bicameral' mind that allowed for right-to-left dictation. The theory I propose retains elements of the Jaynesian model, along with recent evidence that not only the ancients, whose minds we can read in their art, carvings and poetry, but also modern era poets had *atypical* brain lateralization with right-hemisphere enhanced minds.

Much of Jaynes's theory relied on the perception that language itself is a metaphorical process based on our bodies in the world, a view later theorists have affirmed. We do, in fact, speak and act largely without conscious premeditation: if consciousness intrudes *too* much, we can do neither. As Jaynes said, 'Our minds work much faster than consciousness can keep up with' (Jaynes 1976/1990: 42). More specifically, Lakoff and Johnson (1999) said:

> Conscious thought is the tip of an enormous iceberg. It is the rule of thumb among cognitive scientists that unconscious thought is 95 percent of all thought—and that may be a serious underestimate. Moreover, the 95 percent below the surface of conscious awareness shapes and structures all conscious thought. If the cognitive unconscious were not there doing this shaping, there could be no conscious thought. (Lakoff and Johnson 1999: 13)

Undetected mental processing can erupt with a fully formed answer to a difficult problem. While hard forethought has often prepared the way along meandering mental pathways, the response comes when the flow of conscious chatter in the mind has been stilled. Jaynes and others have cited the example of French mathematician Henri Poincaré who, forgetting about his work while on a geological excursion, stepped up on a bus and received a mathematical discovery in one bright flash, like Archimedes' probably apocryphal 'Eureka' moment in the bathtub.

The Sound of Music

Springer and Deutsch (1998), mentioning Jaynes, speculated that the misattributed voices of the gods were possibly the beginnings of inner speech we all now hear, but did not explain why they often arose as poetry and song. Both modes of expression are, in fact, linked through the right hemisphere. Jourdain (1997) explained that 'the right-brain auditory cortex focuses on relations between *simultaneous* sounds... the secondary cortex of

the left hemisphere targets the relations between *successions* of sounds...
and the perception of rhythm' (Jourdain 1997: 56-7). The left ear (right
hemisphere) processes melodies; the right ear (left hemisphere) processes
both rhythm and pitch.

Jourdain used biography, as I do with the poets, to press his point that
composers with a genetic predisposition to atypical lateralization — an
enlarged right temporal lobe — could suffer from hypersensitivity to sound,
Hörlust or 'hearing passion':

> The infant Mozart was made sick by loud sounds; Mendelsohn simply cried
> whenever he heard music. As a child, Tchaikovsky was supposedly found
> weeping in bed, wailing, 'This music! It is here in my head. Save me from it'
> … Handel would not enter a concert hall until after the instruments had
> been tuned, and Bach would fly into a rage upon hearing wrong notes.
> (Jourdain 1997: 188)

Hypersensitivity to sound can translate into superior musical abilities but
with emotional deficits. Autistic musical savants, Jourdain said, have
perfect pitch and a great feel for harmony, but can lack a sense of rhythm.
They are often blind, due to premature birth with oxygen deprivation that
causes extensive damage in the left hemisphere. But the flip side is compensatory musical prowess. The English musical savant, Derek Paravicini,
who has both the passion and the pitch, can distinguish the simultaneous
sounds of a full orchestra distilled into basic chords using his highly
developed power for pitch discrimination.

Jourdain made a very interesting observation about the earliest human
music based on archeological research done in Southwestern France. Here,
the most heavily painted caves were also the most resonant for sound,
suggesting that art and music coexisted in a ritual space, 'accompanied by
flutes and drums and whistles' (Jourdain 1997: 305). Song may have predated speech or have been held in higher regard. The oldest known flute,
made from animal bone with five finger holes, dating back 42,000 years,
was found in a German cave in the Upper Danube region.[1] Archeologists
Chazan and Horwitz (in Bahrami 2014) investigated a much older site, the
Wonderwerk Cave in South Africa, from around 180,000 years ago. Here,
they found 'non-native, non-utilitarian' stones had been brought deep into
the cave where water dripped rhythmically. They suggest, 'a dawning
appreciation of light, sound, and tactile beauty contributed to our cognitive
development' as a species (in Bahrami 2014: 34).

The wedding of art and blooming cognition makes sense. The earliest
signs of human invention, other than those geared solely to survival (tools,

---

[1] Available   at   http://www.nytimes.com/2012/05/29/science/oldest-
musical-instruments-are-even-older-than-first-thought.html [accessed 14
February 2014].

for example), most likely arose from sense impressions in the environment that were converted into lasting constructs binding a small group around a campfire. Damasio (2010) concluded that:

> [T]he arts prevailed in evolution because they had survival value and contributed to the development of the notion of well-being. They helped cement social groups and promote social organization; they assisted with communication; they compensated for emotional imbalances caused by fear, anger, desire and grief; and they probably inaugurated the long process of establishing external records of cultural life, as suggested by Chauvet and Lascaux. (Damasio 2010: 296)

Far from what Pinker called 'auditory cheesecake' (Pinker 1997: 534), music *is* art. All of the arts were and remain a 'homeostatic compensation' and 'remarkable gifts of consciousness to humans' (Damasio 2010: 296).

Jungian psychologist Erich Neumann (1959/1974) said early humans lived in a world where the unconscious and the transpersonal outweighed consciousness and individuality. Indeed, I would say, with less developed frontal lobes for linguistic thought, and a more dominant right hemisphere for surviving in their hostile environment, they would have more easily attained altered states of consciousness with transformative and magical powers, including the ability to make art. Nonetheless, Neumann says it was the 'Great Individuals' within the group who would have initiated the creative impulse, while attributing their knowledge 'to the spirits of their ancestors, to the totem'. Sensed presences, in other words, inspired them (Neumann 1959/1974: 84). These were the shamans whose style of art, dance, music and ritual would confer group identity. Later, poets, claiming divine inspiration, would tell, or rather sing, their epic tales, using their voices to bind the group.

Early Realistic Art and Autistic Savant Skills

Nicholas Humphrey (1999) theorized that, in the absence of language skills, a very young Ukrainian girl living in the UK, Nadia Chomyn, developed exceptional drawing skills closely resembling the realism of cave art and even Leonardo da Vinci's drawings. To support his argument, Humphrey cited Nadia's loss of artistic ability after acquiring basic language skills through intensive teaching. Commentators on his hypothesis countered with provisos. UK artist Stephen Wiltshire, also a mute autistic child with exceptional talent for drawing, did learn to talk while maintaining his artistic ability, cultivated by early teachers and later in art school.

More recent research into autism shows that the language-or-naturalistic art hypothesis does not suffice, given the different forms this neurodevelopmental disorder can take. First, only 10% of people with autism will have special talents, which cluster around calendar calculating; mathematical, mechanical or linguistic skills; art, most often linear perspective drawing, like Stephen's; or piano music played with perfect pitch, like Derek's.

In an update of her own work, Lorna Selfe (2011) said that Nadia drew quickly and excitedly on any available surface when pictures by professional artists inspired her. Selfe theorized that Nadia could execute her subjects with absolute realism because, free of *conceptual* prejudices, she was able to *perceive* with unclouded vision. A virtually mute Nadia began drawing perfectly orchestrated horse and rider drawings at 3½. At 5 ½, she drew an excellent horse whose rider had an odd bug-eyed stare. Aged 6–7, she detailed human feet and shoes without the upper body. As an 8–9 year-old child, Nadia's talent was mixed. Alongside a realistic horse, she drew a simplistic girl, which Selfe believed imitated how other children her age were drawing. At age 11, Nadia drew a horse and jockey (from the rear) copying Toulouse-Lautrec's *The Jockey*, without the surrounding context, but still using good form. After age fifteen, her talent was gone.

Ramachandran (1998) theorized that Nadia may have had 'unusual wiring', with a highly developed right angular gyrus, since 'damage to the *right* parietal cortex, where the angular gyrus is located, can profoundly disrupt artistic skills (just as damage to the left disrupts calculation)' (Ramachandran 1998: 196). He does not discuss the loss of her skill, but it is possible that neuronal pruning may have nullified synaptic overconnectivity for the visuospatial that had compensated for her verbal defect. Selfe says Nadia's talent was exceptional, even among autistics, given the early onset of her drawing, her use of perspective, the depth of her cognitive deficits and the eventual total loss of her ability. Now forty, Nadia is severely learning disabled, neither drawing nor making eye contact with people. Sadly, she is both artless and mute.

Humphrey correctly remarked that Nadia's animals were shown realistically leaping and overlapping like those found in the caves. Both the child's and the Paleolithic artist's were painted or drawn with great dexterity. The rare human in Chauvet is indeed partial, including the lower half of a woman's body, whose left leg is conflated with that of a bison-headed sorcerer, and a fallen bird-beaked stick figure in a state of sexual arousal. In the cave of *Les Trois Frères*, an antlered animal-human, paws raised together, dances rhythmically on human feet.

Experts agree that these figures are shamans who were also the artists. Michael Winkelman (2010) says the wounded birdman in Chauvet represents the shaman's traumatic induction into his visionary world. Modern-day shamans still ingest plants to attain altered states of consciousness in which they merge with animals or receive instruction from plant teachers to bind the community and to promote healing. Intuition, telepathy and prophecy are commonly claimed and confirmed. Similarly, archeologist David S. Whitley, who studies living Native American shamans as well as cave art in prehistoric French sites, agrees on the sacred, ceremonial function of the art in the deep, dark recesses of caves favoring hallucinatory experience. Indeed, conflated animal/human imagery portrayed the shaman's experiences of death, bodily transformation, intense anger, sexual arousal and mystical (out-of-body) flight. For Whitley, the 'half-human and half-animal' drawings were 'embodied metaphors' for the *actual* trance sensation of merging with the animals; however, they were not drawn while in an altered state. The art was meant to provoke fear and confirm the shaman's role as 'master of the spirits' (Whitley 2009: 206).

Certainly, the drug and the dosage make a difference. At their extremes, current-day studies show that low-dose psilocybin treatments provoke insights and personal change at a highly therapeutic level (see Carhart-Harris 2012). On the other hand, when Don José Campos (2011), a shaman living in Peru, drinks the powerful hallucinatory brew Ayahuasca, he hears voices, sees spirits, sings songs and heals others, alone or in a group, gauging intuitively the dosage they need to participate in the ceremony.[2] He reports that there is no time, no space under the brew's influence. Ayahuasca-inspired paintings by Campos's friend, Pablo Amaringo, himself a former shaman, show great complexity, intense color, an inner world of merged or non-merged animals and humans performing ritual practices, either in the wild diversity of nature or in intricately decorative buildings. Their paintings are beautiful, but far from realistic. Amaringo says the brew is effective 'for the artist, for the musician, for the scientist, for the sculptor, for the mystic, for the religious person' (Campos 2011: 52). Clearly, the drug inspires unfathomable creativity, insight, telepathy and permeable barriers between the knower and the known, the seer and the seen; but the imagery represents another realm of transcendent experience, not this one.

Oliver Sacks' *Hallucinations* (2012), suggests that all folklore, religion and aboriginal art derived their myths and imagery from hallucinations. Relying on his own and his patients' experiences, he demonstrates how the quantity and strength of the drug taken as well as the cerebral strengths, weaknesses and psychological proclivities of the experiencer will affect their anomalous sense output. In or out of a drug-induced altered state, a shaman can access spirit 'helpers' with seemingly supernatural powers for good or for evil, find misplaced objects or lost souls, predict where to find game and, especially, heal.

Maraldi and Krippner (2013) say that artistic creativity has been connected to *dissociative* states from the time of shamanistic rock art to modern

---

2   Don José Campos also says 'the prevalence of sexual imagery, dismemberment and rebirth' in shamanic work represents a powerful restructuring and freeing of the self (Campos 2011: 106).

era automatisms. Their study focuses on the automatic paintings of a present-day Brazilian artist, Jacques Andrade, who, pursuant to his spiritualist community singing and praying to call forth the spirits, channels famous artists from the past, creating two paintings at a time using both fists with 'extraordinary speed and dexterity', without feeling fully in control of his movements. The paintings do not resemble the claimed artist's, whose name he imitatively signs below, and are lacking in 'depth and symbolic richness'. Andrade scored very high on both the Dissociative Experiences Scale and the Childhood Trauma Questionnaire. Skin conductance testing showed a 'greater sympathetic activation in his left hand [right hemisphere] than in his right hand [left hemisphere]' and generally incongruent physiological reactions, as in low arousal when processing fearful material, 'not unusual' in spiritual practitioners and channelers (Maradi and Krippner 2013: 554-7).

One can almost imagine a shaman similarly painting in a dissociative state during a ritual meant to impress the group. Others would have had to hold up the firelight and the scaffolding for cave ceilings. Maradi and Krippner conclude with a study by Peres *et al.* (2012) that demonstrates how experienced mediums who practice automatic handwriting, supposedly under the control of spirits, show *less* blood flow in the frontal attention system, and actually write *better* in the trance state than in a control condition. Less experienced mediums showed *more* activation in the same region, and felt as if they were receiving dictation. So, a shutdown rather than an enhancement brought their dissociative talent to the fore.

In another twist, Whitley and Whitley (in press) link madness and creativity to explain the shaman's art, citing accumulating research that genetic mutations introduced mental illness into the human species around the same time as cave art appeared 40–50,000 years ago. The genetic mutations for autism, schizophrenia, bipolar disorder and even ADHD all arose around the same time, i.e. during the migration out of Africa, with increasingly large populations and different environmental pressures making evolutionary change progress more quickly than ever before.

Summing up, Whitley says: 'We became "modern" humans, from this perspective, not when our full rationality alone emerged, but when our full emotional range—including mental sickness—developed' (Whitley 2009: 245). *People with the mutated genes would be hypersensitive to sounds, sights and unusual perceptions needing a creative interpretation to be understood and processed.*

I would add that *beyond* the genes for emotional volatility there were then, as now, traumatic initiations and ritual practices that reinforced the shamanic calling. Combining the two, the shaman would have been more apt to sense spirits and project imagery onto the contours of the rock face. We know from contemporary research that people with a highly active right hemisphere are more likely to claim clairvoyance, telepathy, dream precognition or project faces onto unlikely places. One would think that left-handers would be more prone to magical ideation; but, in fact, people with mixed-dominance claim more anomalous experiences and are more prone to psychosis of the schizophrenic or bipolar variety (Sommer and Kahn 2009). So left-handedness itself would not produce mental illness or a developmental disorder.

French researchers Faurie and Raymond (2004) agree that the shamans themselves were the artists who made their marks. Specifically interested in handedness, they compared negative hand paintings of Paleolithic artists to those of student recruits who blew ink through a special pen onto their outstretched non-dominant hand. They concluded that the proportion of right- to left-handedness had not changed over 10,000 years: 90% right-handers to 10% left-handers.

It seems in the case of cave artists, autistic savants and mediums their minds *needed* to be altered, whether through dreams, drugs, disease, developmental disorders or dissociative techniques, to access exceptional skills. The key may be in the dopamine and serotonin neurotransmitter systems, differentially activated in the left and right hemispheres (Previc 2009). As to shamanism in particular, Winkelman (2010) cites EEG studies on altered states of consciousness that show left/right-hemispheric synchronizing patterns in the frontal cortex as well as bottom-up limbic–cortical synchronization. A loss of serotonin inhibition over dopamine, he says, allows for emotional flooding, ecstasy and hyperactivity of the visual system, elevating non-word-based symbolism into consciousness.

***

Taking past and present together, I propose that Nadia and Stephen, as well as other autists and artists, have atypical, but differing, lateralization from the norm of left-hemispheric dominance for language and right-hemispheric dominance for the visuospatial. Handedness may be significant. The right-handed Stephen may have reversed lateralization: right dominance for language, hence his eventual development of normal language function, and left-dominance for the visuospatial. Enhanced visuospatial memory for literal detail would explain his exceptional skill in drawing the skylines of major world cities. He also has perfect pitch for music, a known left-hemispheric specialty (Goldberger 2001).

The left-handed Nadia may be right dominant for language, already a 30% possibility, as well as for the visuospatial. Selfe believed the left defect/right compensation thesis did not work in Nadia's case because an early EEG indicated unusual activity in the right hemisphere. Functional MRIs were not available at the time and Nadia would have had to produce language to indicate its hemispheric provenance. Selfe reported that Nadia verbalized less when she was under stress, which suggests to me an emotional right-hemispheric interference. Nadia also became increasingly anxious and aggressive as she aged and lost her artistic ability. It may be that art had had a calming effect on her, in line with Damasio's homeostatic thesis. Left-hemispheric aggressivity and hand flapping began when her talent waned.

Differing Examples of Language versus Art

Nadia and Stephen were only two cases of the 'language versus art' model. Many more exist. For example, an American autistic boy, Owen Suskind, who initially spoke, suffered regressive autism. Losing the words, he drew animated characters from Disney films, depicting their posturing, emotions and mouths. When his very caring parents realized he could express himself *through* the characters, they and a series of specialists used 'affinity therapy' to teach him to talk. The Disney characters worked well because they were broadly outlined, mostly animal characters, brightly colored and accompanied by songs, suggesting a right-hemispheric attraction. Nadia too learned better with visual prompts, according to Selfe, but did not have the intensive therapy that worked so well for Owen.

American autistic savant Temple Grandin, whose mother encouraged both her art and her academic success, earned a PhD in Animal Science, designed a body-squeezing machine that calmed down autists like her, as well as livestock handling facilities that did not traumatize the animals. In *Thinking in Pictures*, we see her highly linear, precise, perspectival drawings, which she could visualize in action, with calming curves for the cows to run through. She now writes reviews and introductions in books about others with autistic skills. In her recent review of a book written by a nonverbal child with autism, Grandin (2014) distinguished those who are locked inside their ill-functioning bodies with screaming sensory systems, but remain mentally whole. While unable to speak, they have inner language and can type out their thoughts or point to letters on a grid to communicate. Essayist, poet and scholar Ralph Savarese (2008) explained how his autistic adopted son, who had been sexually abused by a foster family, could express himself in poetry through a talking computer. Not art, but right-hemispheric language emerged silently from his traumatized, neurodiverse mind.

Then there are those who, while fully functional linguistically, even hyperlexic, often lack appropriate neural connections for social relatedness. American professor Priscilla Gilman (2011) described her own autistic child, Benjamin, in *The Anti-Romantic Child*. At 2 years old, he could read poetry, even Shakespeare, plus his mother's dissertation on Wordsworth. But, he could not express himself in spoken language other than by using echolalia or by identifying with characters in books he had read, similarly to Owen. He also had difficulties in body control. As time progressed, Benjamin began reading encyclopedias and nonfiction. Hyperlexic with right-hemispheric deficits, Benjamin read at a highly advanced level, but lacked social skills. Some motor skills, such as chewing and going up and down steps, were difficult as well. Supernormal with details, he did not get their overall context. He recited facts with extraordinary recall, but could not relate to imaginative play. He insisted on the literal at all times and could not understand metaphor or make novel sentences. Nor could he express his sense of self in language.

On the other hand, Benjamin was so hypersensitive to the environment that loud sounds, textures and tastes could overwhelm him and make him panic. The newest research on autism suggests a lack of neuronal pruning causes these hypersensitivities, but may compensate with savant skills (Belluck 2014). Benjamin has progressed enormously with a loving mother and specialized schools. With his super-functioning left hemisphere, he has perfect pitch and is an excellent musician. In fourth grade, he surprisingly remembered Robert Frost poems from toddlerhood, using their rhythm and vocabulary to produce his own haikus and rhymed poems. Counting syllables and rhyme are both left-hemispheric features of poetry (Kane 2004).

Christopher Taylor can read upside down or sideways. Although he lives in an institution for the mentally challenged, he has learned 20 languages from reading books and newspapers, as well as from real life experiences. Kim Peek was born without a corpus callosum. He could read two pages of a book at once, one with the left eye and the other with the right, in 8–10 seconds (Ammari 2011). With language skills apparently spread across his entire brain, he could read a book upside down or sideways. He was the real life 'Rain Man' with his super facility with numbers. He is also a calendar calculator. Similarly, Jacob Barnett could not speak or tie his shoes as a small child. His mother removed him from school and traditional therapy, encouraging him to do whatever interested him. Now a teenager, Jacob is working on a Master's degree at the Perimeter Institute for Theoretical Physics in Waterloo, Canada, and fully functional linguistically.[3]

Daniel Tammet is another high-functioning autistic savant who developed an incredible talent for calculating math and learning foreign languages after suffering epilepsy as a child. He can determine π past the range of a calculator using his synesthetic ability, creating distinct forms and colors for the numbers seen in his brain. According to new research (Ware 2014), the incidence of synesthesia, which creates unusual cross-talk

---

[3] Jacob's Tedx Teen talk (2012) is available at http://www.youtube.com/watch?v=Uq-FOOQ1TpE.

between the senses, was found to be three times higher in autistic subjects than in neurotypicals in a sample of around 300 people. Watching both Tammet and Peek in a video,[4] we see they both write with their right hand. However, both wear their watch on their right arm like a left-hander. Further, Tammet gesticulates with the left, places his poker chips with the left and walks left foot first. Both men show mixed dominance. Tammet speaks 11 languages and learned Icelandic in a week, conversing with native speakers on a national TV station to prove it. It is possible that damage to the left hemisphere during his seizures shifted Tammet's language and numerical functions to the right hemisphere, which, in combination with its natural imagistic ability, enhanced by synesthesia, gave him a largely intuitive way of calculating math and learning foreign languages.

Epileptic children often shift language dominance from left to right and many autistic children also have epilepsy. A savant's ability to learn difficult foreign languages in short order may be made possible through the prosodic specialty of an enhanced right hemisphere, whether attained through brain damage or a genetic shift to right- or bilateral-language dominance.

Exceptional skills can also be acquired through brain injury. After a brutal beating and blow to the back of the head at the age of 31, Jason Padgett became a geometry wizard, literally seeing everything in his environment with colorful visual imagery superimposed. He then plotted out the underlying mathematics of π and fractals. He set out to capture everything he saw in *perfect* drawings. Unfortunately, along with his mathematical and synesthetic gifts, he was initially plagued with crippling obsessive-compulsive disorder (OCD), a left-hemispheric attempt to calm a frightened right hemisphere. After seeing the BBC documentary on Daniel Tammet, Padgett understood his own mental transformation and calmed down. An fMRI and transcranial magnetic stimulation (TMS) in a Finnish lab pinpointed his complex mathematical skill in the left parietal lobe, where it belongs typically. His visuospatial artistic ability may have migrated leftward with the blow. Padgett now sells his work as a practicing artist similarly to Stephen Wiltshire. Both draw with intricate linear precision, but the mathematically gifted Padgett draws lines radiating from circular centers, as he literally sees the world that way.

Atypical Lateralization

Most theorists do not take into account atypical lateralization for language or for the visuospatial. Iain McGilchrist (2009) explained this lack in the

---

[4]   See https://www.youtube.com/watch?v=Uv1E_02ir_c. Accessed 19 March 2015.

introduction to his exhaustive study of the varying functions of the left and
right hemispheres:

> ...[O]nly 5 per cent of the population overall... are known not to lateralise
> for speech in the left hemisphere. Of these, some might have a simple
> inversion of the hemispheres, with everything that normally happens in the
> right hemisphere happening in the left, and vice versa; there is little signifi-
> cance in this, from the point of view of the book, except that throughout one
> would have to read 'right' for 'left', and 'left' for 'right'. It is only the third
> group, who it has been posited, may be truly different in their cerebral
> lateralisation: a subset of left-handers, as well as some people with other
> conditions, irrespective of handedness, such as, probably, schizophrenia and
> dyslexia, and possibly conditions such as schizotypy, some forms of autism,
> Asperger's syndrome and some 'savant' conditions, who may have a partial
> inversion of the standard pattern, leading to brain functions being lateralised
> in unconventional combinations. For them the normal partitioning of
> functions break down. This may confer special benefits, or lead to dis-
> advantages, in the carrying out of different activities. (12)

Both Chris McManus (2002) and McGilchrist believe that these genetic
variations, potentially dangerous for an individual mind or for procreation,
might continue to be passed on in the general population for their creative
potential, as we saw earlier with the connection to cave art. Recent neuroscience (Schneps 2014) also shows the visuospatial advantages of the
language deficit in dyslexia pointing, in my opinion, to enhanced right-
hemispheric processing. As an example, nineteenth-century French
sculptor Auguste Rodin could barely read or write by the age of 14, but
had started drawing at an early age. Inspired by Michelangelo's prints, he
devoted himself to art, attending *L'Ecole Impériale de Dessin*. He could stare
at paintings in museums by day and paint them from memory at night.
Sculpture enraptured him. His reading skill developed with a penchant for
poets, including Homer, Virgil, Hugo, Musset and Lamartine. Several
figures in his famous *La Porte de l'Enfer* (*The Gates of Hell*) were drawn
directly from Dante's *La Divina Commedia* (*The Divine Comedy*). Poetry and
art, both right hemispheric, were entwined. Rodin designed the magnifi-
cent sculpture as a 3-dimensional representation of Dante's work, but
surpassed the plan, adding influences from Baudelaire's *Les Fleurs du Mal*
(*The Flowers of Evil*) (see Pinet 1992).

In a possibly schizophrenic example (Bright 2012), right-handed
Augustin Lesage created art aided by auditory hallucinations. First, while
working in the mines, Lesage heard a voice telling him he would become a
painter — a message spirits of the 'dead' corroborated in séances. His voice-
assisted paintings displayed obsessively fine-detailed, highly symmetrical
linear art, suggesting an extremely bilateral brain.

Humphrey (1999) got it right when he said that highly realistic cave art
disappeared at the end of the Ice Age. Five millennia later in Assyria and
Egypt, the style, as he said, was 'much more conventionally childish,

stereotyped and stiff… Maybe in the end, the loss of naturalistic painting was the price that had to be paid for the coming of poetry. Human beings could have Chauvet or the *Epic of Gilgamesh* but they could not have both' (Humphrey 1999: 122–3). However, poetry, like art, can take diverse forms, depending on which hemispheric esthetic leads the way, sometimes determined by a forerunner in the field, then emulated. Quite possibly, the art of the Babylonian period with its rigidity and high symmetry represented a shift to a new esthetic, along with a more bilateral brain organization; whereas, before, the right had reigned.

Metaphoric Dichotomies:
Left/Right, Evil/Good, Female/Male, Human/Divine

Believing the right hemisphere was the 'god' side of the brain, Jaynes said logically that '[t]he left hand, [was] in a sense the hand of the gods' (Jaynes 1976/1990: 119). Yet, from the Old Testament to the New Testament, 'being on the right hand of God' was considered the favorable position. The left hand is likewise vilified in the Koran. On Judgment Day, those standing to the left of God will be flung into the fiery pit of Hell and made to drink scalding water; those on the right go to Paradise with running streams and beautiful houris. Hot and cold, wet and dry, are contrasting metaphors for good and evil eternal abodes. Of course, the revealers of these early religions would not have known that the right hemisphere controlled the left hand and the left hemisphere the right.

The original 'right' preference may not even have arisen from dominant handedness, but from the fact that the sun rose in the east. The Mesopotamian and Hebrew words for the sun were nearly identical: 'shamash' versus 'shemesh' (Cohn 2014: private communication). Interestingly, even earlier roots show a connection to the word 'shaman' (Winkelman 2010). In *Gilgamesh*, Shamash is the sun god.

But the earliest peoples worshipped goddesses, identified with the moon. Women's monthly cycles and the lunar phases may have, in part, determined these associations. During the Upper Paleolithic, art was natural and rounded, depicting static Mother goddess figures with large breasts, abdomens and thighs, feet coming to a point for positioning as votives in the ground. The head was small, featureless and seemingly thoughtless. The womb and vulvas, sometimes with plants protruding, were also rendered in carvings and figures. The feminine principle in its fullness was metaphorically equated with the generation of all life. Whereas the carved Goddess of Laussel, dating back to 22,000–18,000 BCE, holds up a bull's horn shaped like a crescent moon in her right hand, Astarte from second-century BCE Babylon has a moon attached to her head, a small waist, and her *left* hand is outstretched, connoting something other than fertility.

The role of men's semen would not be expressed until literacy with the Hebrew Bible, the Ancient Greeks and later the Koran. Notably, early Islamic scholars said that Allah had created the pen. All decoration, thereafter, would take the form of extraordinary lettering with no pictorial imagery other than geometric forms. Leonard Shlain (1998), a vascular surgeon, not a neuroscientist, believed, like Jaynes, that alphabet literacy changed the way human minds worked. He connected this change not only to increased left dominance but also to a denigration of women: 'when a critical mass of people within a society acquire literacy, especially alphabet literacy, left hemispheric modes of thought are reinforced at the expense of right hemispheric ones, which manifests as a decline in the status of images, women's rights, and goddess worship' (viii). Shlain marked the downfall of the Great Mother Goddess with the worship of the Word.

The Epic of Gilgamesh

In the Bronze Age (ca. 2350–2150 BCE), we find this first recorded mythic story, carved on tablets in cuneiform writing, but most likely derived from a prior Sumerian version, itself derived from oral inspiration. Gilgamesh reminds his 'brother' Enkidu, formed from a clot of clay by a goddess, not through a womb, how he was allowed to 'rest on a royal bed and recline on a couch at his left hand', a rare positive connotation (Sanders 1960/1972: 91).

In Akkadian art, stone figures are, as Humphrey noted, carved stereo-typically and simplistically, with vertical linear orientation in bilaterally symmetrical scenes. Both hands are used, especially raised up in supplica-tion to the gods, or outwardly, holding up vanquished lions. The mythic hero stands on two hybrid beasts, with the head of the king and the winged body of a lion. Concept and style have overridden the real. Previously, goddesses had stood on lions. Ancient Egyptian art was linear as well, and consistently depicted symmetrical handedness.

In an era when writing would have been extremely rare and picto-graphic at that, limited to the scribes in the priest caste who chiseled using both hands, leaving one's name in stone for future generations was the only means to cheat death. The core of *Gilgamesh* is, in fact, a heroic journey to attain immortality, with Enkidu's death along the way propelling the search. N.K. Sanders (1960/1972) wrote, 'Gilgamesh is the one human character of heroic stature who has survived, though heroic fragments may be embedded in other material, as the "Song of Deborah" is set in the Book of Judges' (Sanders 1960/1972: 20). I mention this because, unusually, a female prophet directs a victory with commands she hears from the God of Israel. In her 'song', a poetic commemoration like most epic stories, she picks up a hammer in her right hand and plunges a tent peg through the sleeping enemy commander's temple into the ground. Now, a woman who kills, rather than gives birth, is heroic. Fertility goddesses have morphed into love and war goddesses. Athena is born from Zeus's head. New myths tell stories to justify practices imposed by a patriarchal leadership.

The Iliad and the Odyssey

Similarly in the *Iliad*, the Goddess muse inspires war: *Menin adeie Thea*, Of Wrath sing, O Goddess! Jaynes said that the breakdown of the bicameral mind happened around 1250 BCE, between the appearance of the written versions of the *Iliad* and the *Odyssey*. Damasio (2010) also said, '[E]arly human minds, less integrated than ours, easily perceived the broken-down, piecemeal reality of our bodies, as suggested by the words that come to us from Homer' (Damasio 2010: 93). More extreme, Jaynes said they displayed no ego consciousness; rather, body parts were propelled into action when the '*thumos*', located in the center of the chest, dictated action. Modern research gives us another way of interpreting the *thumos*, bringing both theorists closer together. The rising sensation in the chest or midriff would be *felt* before the outburst of words or actions, then attributed to a god. A Greek writer has interpreted *thumos* as intense rage, provoked by 'trauma, grief and/or dishonor'. *Thumos*, first felt, then described metaphorically, or rather metonymically (the part for the whole), is an angry call to 'arms'. Today we would call it a rage attack.

Counterintuitively, modern epileptic research documents a similar rising sensation in the chest as a prelude to 'an intense feeling of bliss, enhanced well-being and heightened self-awareness', first described by Fyodor Dostoevsky, an epileptic. How can active rage, as described in early literature, and divine bliss, as described by the Russian novelist, emanate from the same biological space? Simply put, rage requires left-hemispheric aggressive action unless it is internalized as silent, right-hemispheric 'avoidant' anger. Picard and Craig (2009) hypothesized from their tests on ecstatic epileptic seizures that the neural correlate of consciousness was in the anterior insular cortex hidden behind the temporal region. A dysfunction there, sometimes on the right but more often on the left, was the locus of the epileptic 'aura', the sign that a seizure was imminent. Whether rage or epiphany, the *dissociative* feeling of what was happening would constellate around intense emotion, environmental circumstances, the patient's personal belief system *and* particular brain lateralization.

The Schizophrenic Connection and C.G. Jung

Along with schizophrenic voices, Jaynes saw the split minds of epileptic commissurotomy patients and hypnotized patients in the present as a 'vestigial residue' of early bicameral peoples. Neurological studies of his time indicated slightly more right-hemispheric activation in schizophrenics than in normal subjects. Further, some patients located their good voices in the upper right (coming from the left hemisphere), while the bad ones came from the lower left (right hemisphere).

Drawing on his own experiences, Carl Jung's life and work exemplified this left/right divide. His mother had two personalities, No. 1 and No. 2. He suffered much in early childhood from attachment issues and again, as an adult, with the approach of World War I, his break-up with Freud and an extra-marital relationship with Toni Wolff. He and Wolff literally shared merged dream states while living in his house essentially as a second wife (Shamdasani 2011). His breakdown manifested as voices and visions with the compulsion to write painstaking calligraphy, in both Latin and German, illustrating his words with brightly colored, tightly controlled, symbolic imagery. As beautiful as it was personally significant, his schizoid writing lay down the foundation for his theory of individuation in a process others could emulate as well.

I would say both Jung and his mother (her father as well) were predisposed to mental imbalance through bilateral brain organization, with environmental stressors pushing them over the edge. Jung's use of both right-hemispheric poetic writing and highly symbolic, vertically oriented, left-hemispheric painting helped quell the voices, balance his overactive mind and overcome his illness. Both highly verbal (right) and artistic (left), he retained a dissociative figure, Philemon, with whom he walked and talked in the garden, as Yahweh was said to have done with Adam and Abraham; then with Moses, although shielded behind a terrifying light.

One of Jung's patients provided further evidence of a bilateral, schizophrenic mind. In *Memories, Dreams, Reflections* (1969/1989), written by Aniela Jaffe using Jung's notes, we learn of a patient from the early days of his psychiatric career who heard voices. She described a voice in the middle of the thorax as 'God's voice', reminiscent of the *thumos* (Jung 1969/1989: 126). Her other voices were distributed on *both* sides of her body. The 'divine' voice commanded that Bible chapters be assigned before each therapeutic session, followed by a test. After six years of therapy, the voices 'had retired to the left half of her body, while the right half was free of them' (Jung 1969/1989: 127). Both sides could speak, with more negativity coming from the right hemisphere; but the left hemisphere, focused on reading and reciting, had healed. Perhaps 'bicameral' should be restated as 'bilateral', i.e. separate spheres of influence with differing tones and talents.[5]

---

[5]   McGilchrist (2009) thinks Jaynes had made a significant discovery, but that his 'breakdown' logic was faulty. McGilchrist believes that schizophrenia is a left-hemispheric disorder: a 'sort of misplaced hyper-reflexive self-awareness, and a disengagement from emotion and embodied existence' (261). In

The Dissociative Connection

Canadian psychiatrist Henri Ellenberger (1970) provided many case histories of disturbed patients exhibiting distinct personalities with differing linguistic preferences. A young Italian woman named Elena alternated between an 'overtly psychotic' French-speaking personality and a less-ill Italian one who had no knowledge of the French speaker (Ellenberger 1970: 138). Her psychiatrist, Giovanni Enrico Morselli, 'could make her shift at will from the French to the Italian condition by making her read aloud 50 verses by Dante' (Ellenberger 1970: 139). With the enforced use of her left hemisphere, the girl bolstered her conscious self, recalling suppressed traumatic memories of her father's sexual attacks. The traumatized French-speaking self, I would posit, emanated from the right hemisphere. What seemed like schizophrenia was actually a dissociative defense against sexual abuse.

The newest research about stray negative thoughts in normal, healthy people adds more insight still. From fMRI studies, we are learning that negativity comes from the right hemisphere because of its propensity toward fear and worry hardwired into our brains since Paleolithic times as a survival mechanism. The 'reflective' left hemisphere is geared to the future and happier than the wary right, reacting automatically in the present and worrying about the past. Cohn (2013) pointed out the major difference between the aggressive, punitive voices of the much older Torah and the later messages of grace, love and forgiveness propounded in the New Testament and the Talmud, written contemporaneously. The shift from angry bellicosity to a new reflective, positive tone may help explain this difference.

The Atypical Mind of Poet and Artist William Blake

Arthur Koestler (1964) described the act of creation as unconscious mental processes solving a serious problem redolent of Jaynes's theory. He believed that visionary poets, especially, but also mathematicians, scientists, dreamers and madmen must use a process called 'thinking aside', sounding like right-hemispheric processing. Leaps often occurred during transition states from sleep or other relaxed states where left-hemispheric language and logic were suspended. But Koestler also recognized the bottom-up effect, citing a regression to 'an older and lower level of mental hierarchy' flourishing naturally among children and primitives and resurfacing under pressure in creating minds (Koestler 1964: 168). Current scientific literature describes both the shutdown of left frontal processes freeing lower levels to create versus left- or right-hemispheric damage,

---

other words, there is more language (I would say in both hemispheres) while the right-hemispheric bodily sense of self is breaking down.

which disinhibits the opposite hemisphere, increasing dominance and synaptic connectivity for special skills.

Using Koestler's metaphor, Blake was an artist who sank into unconsciousness to bring up pearls of insight, but failed to re-emerge. English professor Judith Weissman's Jaynesian-inspired treatise, *Of Two Minds: Poets Who Hear Voices* (1993), revealed the trajectory and complexity of Blake's poetry, ultimately deeming him 'a madman and a sexist' (Weissman 1993: 314). His early works, she said, were imitative of 'fraudulent Celtic epics of Ossian' as well as 'cerebral and subversive, sympathetic to political revolution... and opposed to the constraints that Christianity placed on the sexuality of both men and women' (Weissman 1993: 129). She asserted that Blake's ideas on religion were initially consciously deliberated, not divinely dictated, with a message that all religions and cultures are equally valid. Predating Jaynes and Shlain, Blake saw clearly how 'writing spells the end of the divine commands that can dominate a culture'. She hypothesized that Blake had read Homer, learning that 'light replaces the living forms of the gods when human culture is losing its ability to hear commanding voices of the Biblical prophets'. The prophets made 'logical predictions' about what would happen if the 'known ethical code' of the group were broken. They were not clairvoyant. Blake may have believed he was hearing divine voices, but much of what he said could be traced to his predecessor poets: the prophets, Homer, Virgil, Shakespeare and Milton (Weissman 1993: 132–3).

A new twist in his poetry occurred with 'The Visions of the Daughters of Albion' (1793). Here, Weissman says, Blake insisted on men's 'total sexual freedom' and women's 'acquiescence only'; he 'abandoned rationality for auditory hallucination... succumbed to fears of women's jealousy, women's assertiveness, and finally, women's very existence' as independent beings (Weissman 1993: 135). Weissman saw no personal evidence for what instigated this misogynistic shift.

That evidence now exists. Marsha Keith Schuchard (2006/2008), during a 24-hour airport layover in London, ventured to Muswell Hill where she was handed the Armitage-Blake family documents from the Moravian Church Archives (Schuchard 2006/2008: ix–x). Here, she learned of Blake's mother's participation in the Moravian Chapel in Fetter Lane, her influence on her son's ideas and her encouragement of his visions. Schuchard explains how Blake attained altered states through unusual sexual practices forced on his wife Catherine, long the unwilling muse. Those refusals (including the suggestion he take in a concubine) led to his misogyny. His practices also explain the erotic imagery prominently displayed in his art, along with arousal that others later erased or covered up with knickers.

In a meeting of atypical minds, Blake came under the influence of the London Swedenborgian society, whose founder, among many other things, was a clairvoyant voice-hearer who developed his own theory of brain lateralization. Ahead of his time and operating by intuition as much as by scientific analysis, Swedenborg (1888) concluded that the left part of the head allowed 'the intellectual things of Spirits and Angels' to flow in, while evil spirits brought in 'direful phantasies and persuasions' (Swedenborg 1888: 432). First saying, in close approximation to reality, that the left is rational and the right emotional, later he contended that the intellectual enters through the right side of the brain and the left is occupied with our affections, as per 'spirits of the planet Jupiter' (Swedenborg 1888: 435). In any case, Swedenborg's 'ability to hear angelic voices, see visions and know clairvoyantly all came to a sudden end three months before he died, immediately after he suffered a stroke' (Wulff 1997: 338–9, fn. 81). Whereas the spirit voices stopped, *he* could still speak. The voices must have been coming from his non-dominant hemisphere.

What can we infer about Blake's inner mind from the outer signs? Surrounded by visions from childhood until the moment of his death, Blake may well have had a genetic predisposition to atypical bilateral lateralization. His beloved brother, Robert, claimed visions of Moses and Abraham too. Peter Ackroyd (1995) said that Blake was also subject to spontaneous 'eidetic', as if real, imagery, a right-enhanced visual feature. He once experienced the combined sight and sound of spirits from the past in a hallucinatory procession in Westminster Abbey.

Early events and hypersensitivities also proved traumatizing for Blake: a favored younger brother, John; a single humiliating beating; a tyrannical father. William was a solitary and incessant reader with a strong personal myth that consoled and set him apart from others while providing material for his art and poetry. But the 19 year-old Robert's loss cut the deepest. As his brother died, Blake saw his spirit rise up through the ceiling, 'clapping its hands for joy' (Ackroyd 1995: 99). He believed that Robert's spirit continued to converse with him and wrote from his dictation. Blake solved a technical problem using dissociative instructions from his brother, creating an etching technique that combined poetry, painting and engraving into a single artistic process allowing him to publish his own work.

The bilateral thesis explains Blake's alternating mood states; his early and sustained talent in both art and poetry, usually linked; his visionary experiences and extraordinary visual memory; his practice of mirror writing, important for his work as a printmaker; his ease in learning difficult foreign languages, Greek, Latin and Hebrew, later French, then Italian to read Dante; and his ability to enter euphoric trance states requiring synchronization of the hemispheres.

In his art, human figures were not only linear, but also frequently depicted with bilateral handedness resembling Akkadian and Egyptian art. The uniformly clothed figures were extremely linear with little discernable difference in facial features or natural realism. His naked figures, and there are many, featured rounded buttocks, muscular male torsos and explicit sexual characteristics in both sexes. In one drawing, a cathedral inside a woman's vagina clearly expressed his erotico-spiritual fixation.

After her husband's death, Catherine Blake claimed they had daily conversations as he sat across from her. Of course, Blake had taught her his methods. But Mrs. Blake claimed to have known she would marry the poet at their initial meeting. Jung had had the same type of presentiment, heard as a voice, when first meeting his wife Emma. Precognition is commonly claimed in right-enhanced minds.

The childhoods of both Jung and Blake predisposed them to religious fixations. Jung's father was a pastor. Even as a child, his son had bizarre hallucinations expressing resentment of the paternal religion. Raised in a household devoted to Bible reading, Ackroyd said it is not surprising that Blake's 'poetry and painting are imbued with biblical motifs and images; the very curve and cadence of his sentences are derived from the Old Testament' (Ackroyd 1995: 13). Once, reading a book, Blake asked himself, 'Who can paint an angel?'. The Archangel Gabriel declared: 'Michael Angelo could'. Gabriel 'proved' his identity by opening up the roof of the house, ascending into heaven, standing in the sun, and moving the universe (Ackroyd 1995: 202). Upward ascent, also envisioned after his brother's death, is typical of an overly aroused left hemisphere, according to F.H. Previc (2005). But perhaps it depends on one's lateralization. Portraitist Thomas Phillips reported the same conversation with Blake, as he painted him pen in right hand, looking upward and leftward from his right hemisphere. Another sign of his lateralization is the centerline between his brows, seen in this same portrait. According to evolutionary behaviorists Sheehan and Reffell (2013), the centerline indicates a bilateral brain, where both sides operate at will.

Blake also famously claimed his revered predecessor poet, Milton, had descended into his *left* foot, showing right-hemispheric provenance. Schuchard (2006/2008) also connected the left 'great' toe to ecstatic states. The fact that the feet are mapped in the brain adjacent to the genitalia would explain how this type of mystical induction would work. Obviously, the poetry, the drawings, the sexuality and the spirituality were all linked. Blake also held séances with a friend, conjuring the faces of the famous dead and drawing their portraits. Relying on faces he had 'once seen or drawn' (Ackroyd 1995: 350) before demonstrates an extraordinary visual memory for faces, once described as a right-hemispheric feature, but now considered more complex, using both hemispheres.

American psychologist Kay Redfield Jamison (1993) labeled Blake bipolar with psychotic features, citing his 'hallucinations and delusions', 'periods of exaltation and grandiosity', 'melancholy', 'excessive irritability and attacks of rage, suspiciousness, and paranoia' (Jamison 1993: 66). Regarding hemisphericity in bipolars, she noted a series of studies indicating a 'pattern of right-hemispheric (or nondominant) impairment… associated with problems in perception, spatial relations, integration of holistic figures, and complex nonverbal tasks' (Jamison 1993: 244). An impaired visuospatial right might explain the linearly oriented figures in Blake's art, in keeping with the autistic savants whose artistic skill emanated from the left; but bilaterality might better explain their coexistence with naturalistically represented naked figures.

In an appendix to *Touched with Fire*, Jamison listed American, English, French, German and Russian creators of the nineteenth and twentieth century whom she pronounced bipolar, including 83 poets, 41 writers, 30 composers and 41 artists. What Jamison did not say is that visuospatial neglect in the real world can be a poetic boon in the creative world. Right-hemispheric, imaginative, metaphoric language can become enhanced, with a sense of divine presence. Blake insisted he only wrote when commanded by angels and *saw* 'the words fly about the room in all directions' (Ackroyd 1995: 363). Therefore, Blake was a synesthete as well, like some autistic savants with atypical lateralization. Poet and prophet, artist and visionary, angry and blissful, erotic mystic and misogynist, Blake had a bilateral mind, organized like few others.

***

Once we clear the way for an understanding of how the hemispheres, whether through bottom-up expression, synaptic hyperconnectivity, synchronization, injury or disease, produce special genius, we see how creativity blossoms widely outside the formulaic boxes of left versus right. Variety is nature's way. One thing remains constant: the mother of invention is often the woman who encouraged her child's special skill. Her loss as well remains an important influence in the child's internal landscape. Swedenborg, who matriculated at Uppsala University when he was only 11 years old and drew representations of the human brain with autistic linear precision, lost his mother at 8 years old, a number we will see again in this study of atypical minds.

Chapter Two

Sensing Presences: Poetry, Religion and the Right Hemisphere

Poets are damned but they are not blind, they see with the eyes of the angels.
— William Carlos Williams

The road to God is paved with many stones: metaphor, poetry, music, ritual experiences, prayer, and meditative experiences.
— Eugene d'Aquili & Andrew B. Newberg, The Mystical Mind

Language, Emotion and the Hemispheres

In the last chapter, we speculated about art and music arising first in early humans before written language increased cerebral dominance, for most people, in the left hemisphere. We also saw how varying types of atypical lateralization can confer special skills in autistic savants, whether in the left or in the right hemisphere. Early shamans appear to have been a special breed, whether through genetic mutations, conferring mood disorders; traumatic events; or altered states arising through repetitive rituals — drumming, chanting, dancing or drugs — all of which point to an overactive right hemisphere. We now know modern humans with a more bilateral brain organization are especially prone to schizophrenia, bipolar disorder and dissociative states, but they can have exceptional abilities in art and poetry. Epileptic seizures, as we saw in the last chapter, can also confer special skills and alter consciousness.

V.S. Ramachandran and Sandra Blakeslee (1998) distinguished left- and right-hemispheric language in this way: the left hemisphere is responsible for 'speech production and syntactic structure and semantics', while the right controls the 'nuance of metaphor, allegory and ambiguity… crucial for poetry, myth and drama' (Ramachandran and Blakeslee 1998: 133). In their view, our dominant sense of self is held in the left hemisphere, which makes up the story that constitutes our belief system, while the right hemisphere, the 'anomaly detector', plays the devil's advocate looking for inconsistencies and forcing paradigm shifts. Similarly, Michael Gazzaniga (1998) called the left hemisphere 'the interpreter' for constructing a confabulated view of the world as we see it, while the right hemisphere expresses a more realistic view of the world as it is.

Robert Ornstein (1997) differentiated the hemispheres more specifically in his book, *The Right Mind: Making Sense of the Hemispheres*. Along with denotative versus connotative meaning, he added: the reading of technical texts versus stories; using literal language versus metaphor, sarcasm, jokes and irony. Importantly, for him, context (right hemispheric) trumps text (left hemispheric) for full comprehension to occur. Furthermore, the left hemisphere is more vulnerable to strokes than the right and the damage is more deleterious. The left hemisphere is also more at risk in the womb, which leads to a number of disorders. In a sense, the importance Ornstein gives to the right hemisphere predates McGilchrist's, without the latter's major shifts in styles during historical periods, such as the left-leaning Enlightenment and the right-leaning Romanticism.

Significantly, linguistic preferences can shift dramatically in an individual when the hemispheres are out of kilter due to organic problems. Ramachandran and Blakeslee cited the case of a 60 year-old neurologist who was thrilled when he started 'thinking in verse, producing a voluminous outflow of rhyme' after epileptic seizures afflicted his right temporal lobe. They wondered if we all have an inner poet in the right hemisphere who can be pressed into service, perhaps even without having seizures (Ramachandran and Blakeslee 1998: 7).

Kane (2004), however, cited many examples of patients who began to write poetry after *left*-hemispheric damage gave the right hemisphere full rein. She reminds us that poetry need not rhyme. The right hemisphere's associative, visual style favors novel metaphors and similes, whereas the left hemisphere's acoustic, sequential specialty produces rhymes. Confirming Kane's view, Mario F. Mendez (2005) studied a right-handed patient who started writing uncontrollably in rhyme after suffering right-hemispheric epilepsy. Mendez concluded that the compulsively rhyming left hemisphere created couplets when released to do so by the afflicted right. Similarly, schizophrenics create 'clang' alliteration and rhymes: illogical, nonsensical language streaming from an overactive left hemisphere.

Woolacott *et al.* (2014) reported the case of a right-handed woman with transient epileptic amnesia (TEA) who began writing rhyming poems a few months after doctors added a low dose of lamotrigine for seizures to her other medications. It is possible that the epilepsy had impaired her left-hemispheric language and the anti-convulsive brought back a fascination for words and rhyming poetry. Her rhyming word obsession may also have compensated for concomitant right-hemispheric visuospatial deficits brought on by epilepsy.

Christa Forster, a highly creative poet, playwright, singer and former colleague, suffered left-hemispheric lesions during a long bout of fever while studying in Spain. Her early poetry was highly metaphoric, with no rhyming at all. When I asked her to describe her origins as a poet, she responded:

> After the fever, I could 'speak' but I suffered from expressive aphasia. I could speak most confidently when I used metaphor and simile, because these devices allowed me to approximate meaning... After this fever experience, after having to use metaphor and simile, I seemed to gravitate naturally toward writing poetry. I could read, appreciate and write it with confidence of spirit. I count the fever experience, and the ensuing expressive aphasia, as my birth as a poet. Not as a writer, but as a poet. (Personal communication)

Michael Trimble (2007) attributes the discovery of emotional differences between left- and right-hemispheric stroke patients to nineteenth-century French neuropsychiatrist J.L. Luys. Damage to the right hemisphere produced euphoria, to the point of mania, or indifference to the debilitating effects of the stroke (see Trimble 2007: 83–4). Left-hemispheric strokes entrained depression, along with aphasia. Sedating one side of the brain versus the other had the same effect. Tellingly, right-hemispheric poetry arises more often from negative emotion. Although never particularly a poet, one evening, when alone and feeling depressed, I wrote a short poem. The content was non-rhyming, with surprising nature-based metaphors that ostensibly took form without conscious forethought. On the other hand, just recently I wrote a poem to read at my mother's funeral. Although it was a very sad occasion, I made the conscious decision to make the poem uplifting. The result was declarative, fact-based and non-metaphorical, with a lilting rhythm and rhymes that flowed effortlessly.

Ramachandran and Blakeslee specify that excessive electrical firing in the limbic system has strikingly emotional effects. Burning from within, those who suffer these seizures see a divine light and find 'ultimate truth' and 'cosmic significance' in trivial events. 'Philosophical and theological' issues predominate and they feel they are in direct contact with God (Ramachandran and Blakeslee: 1998: 179–80). Temporal lobe epileptics also experience hypergraphia—an obsessive, emotional need to talk or write about their revelations, not necessarily in poetic form. Ramachandran cited the case of an epileptic store manager who gave him hundreds of pages of material he had written after sensing oneness with the divine. The man claimed he could recall during seizure every page of certain books he had read in the past.

The existence of many famous writers and several religious figures considered epileptic also shows a connection, albeit speculative, between over-charged temporal lobes and creative language with spiritual content. The list is long: Kierkegaard, Pascal, Newton, Dante, Molière, Sir Walter Scott, Jonathan Swift, Byron, Shelley, Tennyson, Dickens, Lewis Carroll, Dostoevsky, Tolstoy, Flaubert, Agatha Christie and Truman Capote; Saint Paul, the Prophet Muhammad and Joan of Arc. Dostoevsky claimed he and Muhammad had both seen Paradise through their ecstatic seizures. I would offer the possibility that left-sided seizures might cause a language dominance shift to the right, unaffected side, entraining poetic language. Studies of epileptic children with left-sided lesions show this compensatory shift to atypical language dominance, either on the right or on both sides of the brain (Yuan *et al.* 2006). Kane (2004) also noted studies showing a reversal of normal laterality for language for poets in a manic state.

Trimble favors 'bilateral limbic involvement but with an emphasis on the right, nondominant hemisphere' (Trimble 2007: 174). For him, the sense of an invisible presence shows right parietal lobe involvement. The conflicting data on hypergraphia and epilepsy may reflect atypical language dominance (see Knecht *et al.* 2001, and Helmstaedter 1994). Tellingly, in a study of twenty patients with intractable partial epilepsy, 55% of the left-handers showed right-hemispheric dominance for language as well as 22% of the right-handers (in Sabbah *et al.* 2000). In any case, right-hemispheric language can increase significantly both during epileptic seizure and when recovering from stroke (Taylor and Regard 2003). All of this evidence supports the idea of the right hemisphere's role in the non-rhyming aspects of poetic language.

Hemispheric Synchronization and the Divine

Neuroscientist Michael Persinger has thoroughly investigated the connection between over-charged temporal lobes and presence, poetry and religious ideation. In *The Neuropsychological Bases of God Beliefs*, he proclaimed: '[s]horn of their poetic language, the descriptions of most religious leaders indicate temporal lobe abnormalities' (Persinger 1987: 18). He attributes hallucinations, *déjà vu*, altered body states, religiosity and reincarnation themes to transient electrical activity in the temporal lobes (TLTs). The litany of his sources for TLTs reads like a primer for religious rituals. These include incense; hypoxia induced at high altitudes, whether on a mountain top or in a solo airplane; low-level breathing, as in yoga exercises; hypoglycemia caused by fasting; musical stimulation; repetitive movements, such as dancing, swaying and bobbing, used by Sufi whirling dervishes, in Muslim or Hebrew teaching methods, or in shamanic rituals.

Intensely felt emotional experiences may translate into mystical metaphors. Persinger believes that a separate sense of self imbued with the properties and preoccupations of the right hemisphere exists in all of us; but left-handers, artists and religious believers are more prone to 'alien' intrusions in times of personal crisis or biological stress.

Electromagnetically Induced Presences

Persinger developed an experiment to induce a sensed presence or 'God experience' through electromagnetic stimulation of the temporal lobes. Using his wired helmet or a headband called the 'Octopus' to achieve the effect, Persinger tried his device on various people. When using it on his college students, greater stimulation of the right temporoparietal region or equal stimulation of both sides created more of a sense of presence than those stimulated on the left side alone (Persinger 2002: 533).

Evolutionary biologist and confirmed atheist Richard Dawkins (2003) felt almost nothing at all, even though he had hoped to experience some kind of mystical oneness. Science writer and seeker John Horgan felt energy 'ebb and flow, surflike, through [him]'. While recalling a sensed presence experienced in a sensory-isolation tank in the late 1970s, no current 'demons, ghosts, guardian angels, or deities' penetrated his 'skull-sealed, skeptical brain' (Horgan 2003: 97–8).

Journalist and lapsed Episcopalian Jack Hitt felt he was 'being withdrawn from the envelope of [his] body and set adrift in an infinite existential emptiness' (Hitt 1999). He did not sense God, but vividly recalled two appealing long-lost memories: laughing and singing 'Moon River' with a boyhood friend and getting a glimpse of his girlfriend's breasts. British psychologist Susan Blackmore, with a history of Zen practice and OBEs, donned Persinger's helmet. She felt intense emotion and physical distortion, including anger, fear and the sensation of her leg being stretched. She was convinced at the time that the temporal lobes did play a role in altered states.

Hitt gave the helmet a four on a scale of ten for transcendental experiences. Persinger gave Dawkins a four for temporal lobe sensitivity.

Waking Dream and Hemi-Sync

Clearly, reactions to electromagnetic prodding depend not only on the belief systems of the subjects, but also on their genetic wiring and personal history. Other theorists have used different techniques to alter consciousness that suggest synchronous or bilateral involvement of the temporal lobes. Dr. Gerald Epstein, a practicing psychiatrist and assistant clinical professor of psychiatry at Mount Sinai Medical Center in New York City, presented at a week-long Jungian conference I attended at Annandale-on-the-Hudson, New York. At the end of his session, I asked him if he knew about Persinger's research. He responded that he had just been to

Persinger's lab in Canada and tried out the Octopus himself. Epstein's own waking dream therapy uses visual imagery to refashion a traumatic incident from the past by replacing it with an *imagined* positive memory in its place. After a 3-week regimen, the reconstructed scenario replaces the real one, now released from its formerly painful associations.

An adept of Jewish Kabbalism, Epstein is well versed in mystical experiences and the power of visualization, having studied with the Israeli healer Colette Aboulker-Muscat for nine years. Not surprisingly, Epstein's reaction to the Octopus was intense, as he described it to me. With his hemispheres synchronized, he could neither name things nor see images in his now 'hollow' brain. Feeling his heart might burst, Epstein called an abrupt halt to the experiment. Epstein saw no conflict between a machine that produces a God experience and the 'real' thing. The perception of the divine occurs simultaneously, he said, as a scientifically measurable event *and* as authentic mystical awareness. Persinger, on the other hand, considers all such 'presences' intrusions by an over-aroused right temporal lobe, harking back to Jaynes's bicameral mind.

Epstein recognized the dissociative angle, saying his method brings out 'internal instructors': human, animal, mythical or angelic. They provide advice, answer questions and are always available. Creativity, he says, 'begin[s] to flow from many people undergoing waking dream: books are written, paintings are painted, sculptures wrought, music composed, and poetry written' (Epstein 1981/1992: 146, fn. 9). Note the accent on right-hemispheric domains. Epstein's imagery work resembles Jungian active imagination, but with a less 'analytical, logical framework' and without a mythological, archetypal interpretation (Epstein 1981/1992: 175).

To further my understanding, I attended a seminar on Jungian active imagination. A fellow attendee said he had spent several years as a counselor at the Monroe Institute of Applied Sciences in Virginia. Monroe had been very successful in the radio broadcasting industry until spontaneous out-of-body experiences (OBEs) began at the age of 42. Extensive physical testing led only to a diagnosis of 'minor hallucinatory dysfunction' (Monroe 1985: 4). Rather than seek a cure, Monroe embraced his OBEs as real, beneficial and worth teaching to others.

Monroe invented a method to synchronize the hemispheres through sound. Using headsets, Monroe's 'Hemi-Sync' method funnels binaural beats to people inside isolation booths. Curious professional people trying his method reported having OBEs and encountering seen or unseen intelligences. In their altered states, participants also claimed 'melding' with others or receiving messages from the 'divine' or 'beings' from other 'realities', preparing them for 'multidimensional levels of consciousness' (Monroe 1985: 67). As described, the 'entities' were limited to the vocabulary of the subjects, or, alternatively, used telepathy or direct possession of their vocal cords to communicate verbally (Monroe 1985: 71). Advanced in his own technique, Monroe claimed to receive information by 'rote', an energetic 'thoughtball' that quickly assimilates 'human compressed learning' (Monroe 1985: 161). Swedenborg claimed angels communicated through 'thought balls', which he recorded in automatic handwriting. A life-ling stutterer, he may have had incomplete language dominance, with a bilaterally-organized mind like Blake's or Jung's. Unlike Blake's voices, Swedenborg's began in his fifties, then fell silent after a stroke which took away his own speech. To his delight, the voices returned shortly before his death (see Wulff 1997: 107–8).

Monroe's popular works, *Journeys Out of the Body* and its sequel *Far Journeys*, are collections of stories about these dissociative presences. In an appendix to the latter book, Drs. Tremlow and Gabbard, who did EEG testing on Monroe, noted a shift in high amplitude patterns to the right hemisphere, concluding that he was in a state of deep relaxation resembling a borderline sleep state. They also reportedly observed a wave-like distortion over the top half of his body not easily explained (in Monroe 1985: 273–4).

The Monroe devotee at the Jungian seminar I attended claimed he no longer knew whether he was dreaming or awake, suggesting he was permanently on the borderline of sleep. He claimed to be a shaman who entered the 'void' to bring back missing pieces of his clients' souls to heal them. When I asked him to describe the 'void', he quickly responded with an acerbic tone: 'I can't describe it any more than you can describe the taste of chocolate'. Although he did seem strange, he was not mad. I could not help but wonder how he operated in the world; but, as we will see, others who report extremely altered states of consciousness, where the self is lost, seemingly can.

Contemplation-Inspired Ecstasies

Philosopher-mathematician Charles Musès attained an anomalous state by contemplating a difficult math problem. He described his highly stimu-lated mind as 'lifted to a peak point' where 'logic and imagination are perfectly wed' (in Leonard 1999: 25). This metaphoric wedding brings to mind Persinger's helmet, Epstein's waking dream therapy and Monroe's far journeys, all of which required hemispheric synchronization. Another philosopher-mathematician, Franklin Merrell-Wolff, referred to his own ecstatic states as 'bimodal consciousness', where his relatively conscious self was one with 'unlimited and abstract Space' (Leonard 1999: 65):

> The Godless secular universe vanishes, and in its place there remains none other than an all-enveloping Presence of Divinity itself. So, speaking in the subjective sense, I am all there is, yet at the same time, objectively con-sidered, there is nought but Divinity everywhere... The sublimated object

and the sublimated self are one and the same Reality, and this may be represented by the judgment: 'I am the Divinity'. (66)

Like Musès, Merrell-Wolff had no apparent notion of the hemispheric processes involved in the dissociative state he described. We have only his *words* to confirm an altered state where divinity becomes entangled in the self's perceptive web. His conversion experience brought about this avowal: 'He who knows the Awakening becomes something of a poet, no matter how little he was a poet before. No longer may thought remain purely formal. The poet pioneers, while the intellect systematizes. The one opens the Door, while the other organizes command. The functions are complementary' (in Leonard 1999: 146).

This poetic transformation surprised Merrell-Wolff. But poetry is often the only way that 'ineffable' transcendental states can be expressed in language. As Milton and Blake experienced, writing from an elevated right emboldened by a spiritual presence flows automatically and effortlessly, since conscious, tentative, logical construction is absent.

Drug-induced Dissociation

As we saw in Chapter 1, the dissociative muses of shamans can take many forms as the multiple minions of the right hemisphere. The greater intelligence from the sacred dimension can even present as plants or animals, like those encountered when Amazonian Indians take Ayahuasca.[1] Professor Benny Shanon of the Hebrew University of Jerusalem catalogued the common physical effects and thematic content of Ayahuasca usage. He noted that language is *poetic*, containing 'intense metaphoricity', a 'pervasive metaphorical, and poetic perspective' (Shanon 2002: 243). He even heard a man *spontaneously* recite a lengthy poem about an Egyptian pharaoh. Shanon called the feat 'impressive', but not paranormal (Shanon 2002: 259). Generally, though, the Ayahuasca experience is visual. Verbal content is seemingly telepathic, as in Monroe's OBEs.

The absence of heard speech, coupled with an accent on visual analogies, geometric patterns, art, dance, song and the felt presence of guides, guardians and teachers (usually behind the back), makes the case for increased right-hemispheric activity here too. Elsewhere in his book, Shanon says he and others experienced enhanced performative skills. In quasi-religious language, he claimed they had tapped into a 'cosmic source of plenty' (Shanon 2002: 177); the Muses had descended with 'the coopera-

---

[1]   Schaafsma's (1980) study of North American Native wall carvings shows images of shamans with small beings on their left shoulder, ostensibly speaking into that ear, a possible right-hemispheric connection to hallucinated voices.

tion of superior forces' (Shanon 2002: 220–1). Shanon attributed entheogens ('God-releasing' substances) to the founding of ancient religions.

Psychotic Episode or Spiritual Emergency?

Ayahuasca trance states show similarities to psychotic visions. The same sorts of visual themes and felt sensations Shanon described are present in Russell Shorto's book, *Saints and Madmen* (1999). These include human metamorphosis into power animals; beams of light passing through the body; connecting with plant life; feelings of interconnectedness; seeing auras; heightened awareness of one's place in the universe; and the presence of God and angels. Boundlessness, cosmic significance, deep meaning and energetic light flow are always present, all of which indicate highly aroused temporal lobes.

Shorto tells the tale of David Lukoff, now a clinical psychologist and professor of psychology at Saybrook University in San Francisco. In the early 70s, Lukoff dropped out of Harvard graduate school and took LSD while hitchhiking across the United States. Lukoff (1990–91) tells his own tale in the following quotes from his unpaginated PDF. One evening, he looked in a mirror and saw his right hand glowing with light, with his fingers in the Buddhist mudra position. Lukoff wrote furiously for five days, convinced he was the reincarnation of both the Buddha and Christ and that he had merged minds with the great theorists in the social sciences and humanities. Then, he mailed eight copies of his 'new Bible' or 'Holy Book' to friends and relatives. Having read Suzuki's *Manual of Zen Buddhism* the day before the event, he believed he 'had solved the riddle of Zen teachings and had become "enlightened"'. In right-hemispheric fashion, his 'Bible' contained 'parables, poems, and instructions on how to organize a new society'.

On rereading his work, however, Lukoff recognized his grandiosity. During the isolated, troubled time that followed, he lived alone in his parents' summer cottage. Contemplating suicide, he heard a hallucinated voice command him to '*Become a healer*'. Several years later, while training, then practicing, as a psychologist, Lukoff realized that LSD had triggered his psychotic episode, along with the impact of 'intensive reading on Zen, introverted journal writing, social withdrawal and little sleep'. A full seven years after his 'enlightenment', literally meaning seeing the light, Lukoff went into Jungian analysis and had 'a dream in which a large red book appeared'. Lukoff had somehow visualized what we might surmise was Jung's now famous *Red Book*, recording his own bout with psychotic voices, but not published until 2009, almost ten years after Lukoff's dream. Lukoff also identified with Beat poet Allen Ginsberg, who claimed he had heard the voice of William Blake, as described in his long poem 'Howl'. Not mad, but inspired, Lukoff now *sought out* hallucinatory experiences through chanting and drumming rituals, communicating with trees and the ocean. He learned to control and contain his ecstatic states and 'create a personal mythology'. He became a shamanic healer; and, along with two other psychiatrists, successfully lobbied to have the category 'spiritual emergency' added to the *Diagnostic and Statistical Manual of Mental Disorders* (DSM IV).

Meditation and Unitive States

Andrew Newberg, now director of research at the Jefferson Myrna Brind Center of Integrative Medicine and a physician at Jefferson University Hospital, began writing books on 'neurotheology' when he was a professor in both the radiology and religious studies departments at the University of Pennsylvania. He and the late Eugene d'Aquili used PET, SPECT and fMRIs to get directly into the minds of people in altered states of consciousness.[2] Initially, they studied only Tibetan meditators and praying nuns, not a broader swathe of the population. They saw a spillover effect in the autonomic nervous system when the 'fight or flight response' becomes activated at the same time as the 'quiescent', extremely relaxed, system. Both intense physical or mental activity and slow rituals or meditation could bring on this simultaneous functioning. As they explained it, when confusing sensory information or existential problems overwhelm the mind, mythic stories with gods, spirits and angels have been the long-honored way of making sense and installing cosmic authority.

In their view, overstimulation causes both the right and the left hemispheres to *decrease* neuronal flow in the parietal lobe (which borders the temporal lobe). Since the right is responsible for orienting the body in space and the left for sensing the bodily limits of the self, a feeling of boundless, infinite, timeless space results when those functions shut down.

Modern Mystics

Following the trail of some modern mystics with similar experiences of merger or loss of self may be worthwhile. Suzanne Segal described a unitive state she attained in her aptly named book *Collision with the Infinite* (1996). After practicing transcendental meditation for a number of years, she experienced this dramatic event:

> I lifted my right foot to step up into the bus and collided head-on with an invisible force that entered my awareness like a silently exploding stick of dynamite, blowing the door of my usual consciousness open and off its hinges, splitting me in two. In the gaping space that appeared, what I had

---

[2] Newberg, now collaborating with Mark Waldman, wrote *How God Changes Your Brain* (2012/2013) and *Words Can Change Your Brain* (2009). The latter work uses neuroscience to promote positive thinking.

> previously called 'me' was forcefully pushed out of its usual location inside
> me into a new location that was approximately a foot behind and to the left
> of my head. 'I' was now behind my body looking out at the world without
> using the body's eyes. (Segal 1996: 49)

Altered parietal lobe activity had somehow distorted her body image and
the limits of self, which she later interpreted as a 'bus hit' – a transcendental event in keeping with her Buddhist belief system. Nonetheless,
an empty, frightened self remained, seeking to understand her transformation. Like the Monroe adept discussed earlier, Segal could still
function in the world and was able to get a PhD during this time. In her
search for understanding, she gave up on traditional Western psycho-
therapists, who diagnosed repressed childhood memories, and turned to
Buddhist and Hindu spiritual teachers, who congratulated her on attaining
a 'profound spiritual awakening' (Segal 1996: 112).

The late Ramana Maharishi's work especially inspired her. From him,
she learned that 'one must serve... the unmanifest sat for twelve years in
order to attain Self-realization' (Segal: 1996: 121–2). Assigning special
relevance to the number, she realized she had been in this state for twelve
years. A sustained unity state then occurred while driving in a car:

> ...[S]uddenly I became aware that I was driving through myself. For years
> there had been no self at all, yet here on this road, everything was myself,
> and I was driving through me to arrive where I already was. In essence, I
> was going nowhere because I was everywhere already. The infinite empti-
> ness I knew myself to be was now apparent as the *infinite substance* of every-
> thing I saw. (Segal 1996: 130)

Fifteen years after her anomalous experience had begun, she died from a
massive brain tumor. Although it is unclear when the tumor had begun to
develop, it is likely that, in tandem with her meditative practices, it had
triggered her experience. Meanwhile, she had become a mystical guru
whose followers in California drew spiritual sustenance from her presence
and teachings. When she fell ill and her 'bus hits' had become all too
frequent and debilitating, a flood of childhood abuse memories made her
question whether the entire experience had really been a 'divine'
awakening at all.

American neuroanatomist Jill Bolte Taylor suffered a severe left-hemispheric stroke, leaving her without words, witnessing herself from the outside. She too lost her boundaries and felt a euphoric, expansive sense of
oneness. It took eight years to regain her language and cognitive functions.
In her TED talk and book, *My Stroke of Insight* (1996), we learn how she
interpreted her experience as a rare opportunity for a brain scientist to
observe the brain *from the inside* while perceiving the external world
through a right hemisphere 'liberated' from her left.

Katie Byron entered a sustained transcendental state after having spent many years in a depressed, agoraphobic, paranoid state. She entered a halfway house, confined to the attic, because she was so profoundly troubled. While lying on the floor, she felt, then saw, a cockroach climbing over her foot. In that instant, she shifted into a sustained altered state of consciousness resembling pure mania with a dissociated sense of euphoric being. Despite what seemed like madness, she became an international sensation, with books, workshops and audio recordings helping others to change their lives. When an interviewer asked Katie Byron if her transformation resembled psychologist Abraham Maslow's definition of a transcendental experience, she agreed that it did. In my estimation, she shifted from deep depression to mania, only needing three hours sleep and not eating. She also claimed there was no difference between her dream and wake states. Nothing was real.

Lukoff, Segal, Taylor and Byron all suffered a major shift to their sense of self and cognitive functions through drugs, organic injury to the brain, techniques or mental illness. Yet, their beliefs and practices helped convert mental chaos into a newly ordered reality. With their hemispheres operating outside normal bounds, they could somehow function; and, in fact, their newfound beliefs were transformative in educating and healing others.

Non-Pathological Unitive States

Maslow (1964) had gathered enough evidence from contemporary religious experiences to state that 'lonely prophets' had revealed religions using the 'conceptual, cultural, and linguistic framework' of their time. Further, 'all religions are the same in their essence and have always been the same' (Maslow 1964: 19–20). As we have seen, ordinary people, under exceptional circumstances, can experience the divine when relations between the hemispheres are altered in some way, whether through synchronization, epilepsy, mental illness or organic damage. When these states do occur — Maslow called them 'peak' experiences — people must resort to religious and/or poetic vocabulary to describe them. In a simple formula, Maslow said, 'organized religion can be thought of as an effort to communicate peak-experiences to non-peakers' (Maslow 1964: 24).

While the *internal* experience may feel the same for everyone, the trigger that sets it off, 'for instance in males and females, can be quite different' (Maslow 1964: 29). Interestingly, 'transcendent experiences seem to occur more frequently in people who have rejected an inherited religion and create a new one' (Maslow 1964: 34). Arthur Deikman, a psychologist cited in Ornstein's *The Psychology of Consciousness*, described mystical experience as a 'deautomatization' of habitual perceptual and cognitive distinctions resulting in a merger of all boundaries such that the self is no longer experienced as a separate object. Perception returns to a pure, primitive, heightened mode, without cognition (in Ornstein, 1972/1986: 206–7).

On one occasion, I clicked into a unitive state spontaneously for about fifteen seconds. During the first year I was working on my doctorate, my husband and I took a weekend trip to San Antonio, TX, with our two young sons. Standing in the Spanish Market with my family at a short distance, I was awash in foreign sounds, sights and smells, as music wafted out of brightly colored Mexican storefronts and restaurants. As I calmly licked a bright, blue water ice, I entered a profoundly joyous altered state with no sensed boundary between my inner and the outer world. The *feeling* of my 'mystical' experience emanated from my heart area, like Gerald Epstein in Persinger's Octopus. I saw individual molecules glowing like tiny sparks of light, dotting outwards infinitely beyond me and felt time *literally* stand still. I later termed my experience as feeling like 'nirvana' or 'being one with the divine', as many have before me. I later suspected that the exotic open space, my senses, highly stimulated all at once, and my relaxed state, differing markedly from the intensely verbal experience of my doctoral studies, sparked a brain event in line with Persinger's, Newberg's and Maslow's research.

The contrast between the broadness of my sense stimulation and the precise lick of the water ice may have been the tipping point, like Segal's step on the bus or Katie's cockroach, a bridge to hemispheric synchrony or, possibly, a temporal lobe transient (TLT).[3] Maslow says that people having an experience like mine get 'a sense of the sacred' through 'the momentary, the secular, the worldly' (Maslow 1964: 68). When I told my story in the Jungian active imagination seminar, the instructor said: 'You've had an epiphany? That means you've been called'.

In my study of poets, there usually *is* a reaction against the old religion and a desire to create a better one, entraining a sense of being 'called' on a divine mission. I had no conversion experience, only the desire to regain the sensation. I can recall four precise moments where I felt as though a veil had been lifted in normal perception and I seemed to merge minds or see people as they 'really' were, not as I had perceived them before. These phenomena all occurred during my doctoral days and never since. I was either 'cured' or 'ruined' by a PhD.

When I took a Buddhist meditation class in London, the Western monk teaching the class told me I could get the feeling back any time through meditation. I had many strange physical sensations and several flashes of

---

3   Fellow participants at the 2008 Toward a Science of Consciousness conference suggested my endogenous MDMA, also found in Ayahuasca, had triggered my experience.

self-revelation, but nothing like the San Antonio experience. In fact, meditation had the opposite effect on me. It made me anxious.

I later tried a course with a Tibetan Buddhist in Houston. His guided meditation using metaphoric, visual techniques felt all too concrete and distressing, most likely because of my right-hemisphere dominance. In yet another class, where a highly placed Tibetan lama once visited, class members submitted written questions to him. I asked why meditation made me anxious. Through the interpreter, he said fear was a common reaction in beginning meditators; but, if I worked at it, I would get positive benefits. There *is* evidence that meditation can cause anxiety, even psychosis. Yet, those who are anxious by nature or who suffer transient states of crisis triggered by 'grief and loss' are *more* likely to experience mystical experiences and a sensed presence, attributed to 'hyper-connectionism' of cortical-limbic systems (Persinger 1987b: 191).

In their study of Buddhists and Franciscan nuns, Newberg and Waldman noted increased activity in the frontal lobes ('attention area') of both types of practitioners; greater activity in the right hemisphere of the nuns, focusing on 'the meaning, interpretation and rhythm of speech', where the Buddhists concentrating on sacred images activated the visual processing area in the inferior temporal cortex. Both Buddhists and nuns showed *decreased* activity in the parietal lobes, dubbed the 'orientation area', which suspended their sense of time and space. Finally, and most significantly, as the parietal lobe activity went down, one of the two thalami (left or right), which processes incoming sense impressions, became *more* active. This unusual asymmetry did not appear in people who were more casual about their religious or meditative practices (Newberg and Waldman 2007: 175-6, 183).

Maslow made clear that peak experiences are not necessarily 'ineffable'. Granted, 'sober, cool, analytical' language cannot do the job. However, by learning 'to shift over more and more to figures of speech, metaphors, similes' — poetic speech — the experience was communicable (Maslow 1964: 84-5). Maslow's emphasis on the need for communication and collaboration will play out in the shared practices of the poets we will study.

Phylogenetic Origins

Anomalous happenings produced when the hemispheres work in synchrony must have served some evolutionary advantage to occur at all. Alondra Yvette Oubré believes that mystical awareness probably existed in our most primitive proto-human ancestors, long before the shamans of the Stone Age worked their magic. Gazing fixedly at fire and natural wonders, as well as rhythmic ritual movements and group chanting, would have brought on altered perceptual states. Synchronized movements and synchronized vocalizations would have provoked, in her estimation, a waking dream state full of fantasy, imagery and creativity. More fully developed hominids with speaking capability could have similarly used chanting to promote healing. Left-hemispheric verbal chanting would have calmed the disordered, fearful elements in the right hemisphere. The ritual would have produced synchronization and a euphoric state reducing anxiety, allowing healing to occur. Increasing awareness of death in the developing hominid mind would also have increased the need for rituals to combat grief from loss and separation. Oubré believes that transcendental experiences 'may be rooted in both the genes and cultural constructions that give meaning, reality, and purpose to life' (Oubré 1997: 153). We must also consider psychotropic plants as a source for transcendental experiences.

Shamans and seers would have been the creative individuals whose progressive leaps of awareness made discoveries that advanced our species towards the modern hominid mind. In Oubré's view, 'hominids indeed may have a psychobiological tendency — an innate propensity of sorts — toward attaining altered states, particularly states of transcendental consciousness' (Oubré 1997: 150). The 'enhanced vision' in the altered state would have allowed the 'emergence of symbolic thought' and 'promoted social solidarity among members of early hominid societies engaging in group rituals' (Oubré 1997: 150–2).

Independent scholar Ellen Dissanayake (2000) connected numinous perception with the origins of art, specifically in the intimacy of mothers relating to their babies. Creative adults in early societies, she says, instinctively replicated in dance, music and poetic language the sing-song speech and rhythmic movements of maternal relational activity. Like Oubré, she says that these rhythmic activities created positive emotional states to unite and control the group. Even today, healing songs learned through spirit guides diagnose and cure illnesses. As Dissanayake rightly said, the 'intonation (speech melody), rhythm, stress, tempo, amplitude, pauses, and voice quality' of the mother-to-infant communicative style resurfaces in emotion-laden adult expression (Dissanayake 2000: 46). These prosodic speech qualities are especially evident in poetic expression, governed by the right hemisphere. On the other hand, injury to the right hemisphere can produce monotone speech patterns, devoid of emotion. An impaired right hemisphere also compromises body image, social relations and the ability to control impulsive behavior (Devinsky 2000).

***

In modern, socially fragmented times, disenchantment with the established features of religion leads seekers to find their individual bliss, calling or healing by differing means using the combined skills of the hemispheres in waking dream, active imagination, controlled OBEs, meditation, intense contemplation or psychoactive drugs. Sometimes the 'call' comes unheralded during an intense crisis or, alternatively, in an extremely relaxed state. The new or altered perception can be a surprising, even life-changing, force. Sometimes, a heroic bout with the angels and demons created as a by-product of mental crisis can lead to a true awakening, serving both self and others.

Neuroscientific research and case studies show the inner workings and outward manifestations of an over-aroused right hemisphere, leading to synchronous bilateral activation and a right-hemispheric intrusion in normal consciousness. Bipolar disorder as well as schizotypal traits may increase creativity because of enhanced right-hemispheric dominance, but full-on schizophrenia does not produce the tightly sculptured, sense-drenched, imagistic and rhythmic style of a poetic genius. What else is going on in the mind of the poet?

Chapter Three

Emotion, Dissociation and Linguistic Creativity

Pathologies like brain lesions, epilepsy, bipolar disorder and psychosis can dramatically alter a person's sense of self, thought processes, belief system and language usage. A spontaneous epiphany, contemplation or medita-tion can create a feeling of oneness with the divine. Mind-altering tech-niques can also bring out speaking presences. All of this means that voices and visions, which have always been a part of human experience, are neither supernatural nor necessarily pathological, and are probably the result of similar brain events. Preliterate peoples and young children are more likely to experience imaginative, magical thinking because the right hemisphere is dominant before the advent of writing. Children younger than 10 years old, especially those with more imagination, can confuse 'reality, fantasies, and dreams' due to an overproduction of neurons in the frontal lobes and inadequate separation between the hemispheres (Newberg and Waldman 2007: 119). Dissociation persisting beyond then suggests a predisposition to atypical lateralization and a triggering trauma.

In altered states without heard speech, messages are often conveyed via an image or as seemingly telepathic transference. Names, speech patterns and body language differ in dissociative identity disorder, where alter personalities are given a concreteness they might otherwise not possess. In schizophrenia, both language and discrete body states break down as thought becomes disordered. Neuroscientist and literary scholar Nancy Andreasen (2014 n. pag.) pointed out that James Joyce's daughter was schizophrenic and suggests he was on the spectrum, referring to the 'near-psychotic neologisms and loose associations of *Finnegan's Wake*' to make her point. Yet, perhaps Andreasen did not recognize the poetry in his prose, such as this, in *Ulysses*: 'The air without is impregnated with rain-dew moisture, life essence celestial, glistening on Dublin stone there under starshiny coelum. God's air, the Allfather's air, scintillant circumambient cessile air. Breathe it deep into thee' (Joyce 1922/1992).

Reading, Writing and Rousing the Divine

Just as the ability to sense presences fell along a continuum depending on the temporal lobe lability of those who donned Persinger's electromagnetic helmet, the same is true for people who experience unusual sense or motor events during periods of significant verbal stimulation. In their 'Muse Factor' experiment, Persinger and Makarec (1992) sought to connect the ability to sense presences to creative language usage. Nine hundred college students responded to a questionnaire about their psychic experiences, heightened emotions, feelings of depersonalization, visual, auditory and olfactory anomalies, along with three key questions about a 'preference for creative writing, the capacity to discern profound verbal meaning and the feeling of a presence' (Persinger and Makarec 1992: 219).

The results of their study linked the last three factors in both men and women. The sense of presence was stronger for women and attributed to their 'stronger propensity to experience verbal meaningfulness'. Persinger and Makarec hypothesized that 'strong semantic affect [*an emotional reaction to meaningful words*], as well as novel and unusual combinations of words', would prime the left temporal lobe. Intense electrical firing on the left would stoke 'nonverbal representations' on the right, producing a sensed presence along with a possible message of cosmic significance (Persinger and Makarec 1992: 220). They connected this highly aroused dissociative state to Jaynes's god-side of the bicameral mind, arising in modern times during literary or musical creativity. Readers, writers, ruminators and contemplators would be particularly prone to a sense of presence, depending on how emotionally reactive they were. Sufferers from limbic epilepsy would fall at the extreme high end of the continuum, very creative individuals in the middle, and average people gradated towards the extreme of being unable to express emotions at all.

Their hypothesis, derived by questionnaire rather than by imaging, may only get it partially right. The researchers did not figure in a *reason* for their readers' and writers' intense emotionality in the presence of prose and poetry. Nor did the researchers recognize how strong emotion and the use of poetic devices already demonstrate increased right-hemispheric processing. The right hemisphere's role in creating and interpreting novel or unusual metaphors, first outlined in Kane's (2004) organically based research, has been reconfirmed using fMRIs (Gorana Pobric *et al.* 2008). Nor would left-hemispheric linguistic processing in itself produce a right-hemispheric presence. Rather, if engaged to an intolerable level, the right hemisphere might spin off an 'Other' to facilitate homeostasis through calming words or imagery.

I can provide several personal examples. After an emotional verbal dispute during a trip to Bath, I lay down to sleep. As I closed my eyes, brightly colored paintings flashed by with lava-like rapidity. Although an avid art enthusiast and museum-goer, I did not recognize any of these paintings. Whether created or real, this autonomic artistic flow was clearly meant to calm me down.

This relatively minor event became a full-on encounter with the dissociative, not my own, when a dear friend visited me in Houston after my return from England. Sitting cross-legged in my backyard, she told me how bizarre things had been happening in the nine months since her mother had died. With her head down, swinging back and forth rhythmically, she then introduced me to one of her 'angels' who had materialized in the aftermath of her loss. In the interim, she had been learning meditative techniques and how to channel. This troubling event impelled me to read voraciously in all of the mystical and psychological literature I could find to understand what had happened. One night, I got up at 4:30 a.m. to read a textbook on dissociative identify disorder, then went back to sleep at 6:30 a.m. In an ensuing dream, a mental patient lay stiffly on a psychiatrist's couch, his eyes rolling scarily in his head. Out of his O-shaped, neon-lit mouth, he boomed in a frightening Darth Vader-like voice: 'Freud only got it half right! Read the two Hyperion poems!' With my heart palpitating wildly, I woke up.

As a French literary scholar, I was not familiar enough with English poetry to recognize the poems. However, amongst my books, I found one on English Romanticism. I learned that John Keats had written two poems, 'Hyperion' and the 'Fall of Hyperion', in the two years before his death at 25. I had even read *about* the poems three years earlier in Camille Paglia's Freudian-oriented *Sexual Personae* (1990) and had seen a reference to them in Jungian Marion Woodman's *The Ravaged Bridegroom* (1990). The memories of these readings were so vague that my dream character had not even pronounced Hyperion correctly. In Paglia, Keats was an emasculated poet swamped by a dominating goddess figure. Woodman unveiled the Romantic search for the lost feminine in a 'newly emergent masculine, which Keats identified with the sun god Apollo', and the sad, silent transfer of knowledge from the 'hollow brain' of the goddess Moneta (Woodman 1990: 15).

Now reading the 'Hyperion' poems, along with 'Endymion', for the first time, I discovered the voices of the gods and goddesses, borne by the winds and the waves, or whispered into the ear of their chosen one. Suffering, resonant attraction or a call to service brought them forth. They consoled and illuminated their protégés on the meaning of life and their role in it. Keats's poetry contained a litany of New Age themes long before the 1960s, including the oneness of mysticism, compassion for the suffering, and the acquisition of knowledge by unconventional means. My dream message, which had arisen so powerfully and succinctly, had come from me, not from either theorist. Whether an uncanny synthesis of forgotten readings or a message from a helping other unconsciously created, did it matter? I would never have initiated my research into the minds of poets had it not been for my dream voice.[1] Ted Hughes (1992) wrote that a poet's future self could dictate words to him in the present. Could that have been the case in a literary scholar as well?

Months later, a British friend told me that Jung's autobiography, *Memories, Dreams, Reflections*, contained accounts of his own dialogs with inner figures. Reading this book, I found that Jung's critique of Freud eerily replicated my dream message. Freud had overemphasized the sexual libido in his theory. In such a 'flight from himself, or from the other side of him which might perhaps be called mystical…' we are 'considering only half of the whole' (Jung 1961: 152–3).

Knowing he would be there, I approached Jungian minister and dream therapist Jeremy Taylor in the receiving line of a Unitarian church in Houston. I told him my story and asked where he thought the voice had come from. Pointing his finger at me, he said, 'It came from you… and not you'. Taylor believes that nightmares are designed to wake us up to truths that might otherwise evaporate in the dream mist. He had had his own eerie voice experience while fully awake, recounted in his book, *The Living Labyrinth* (1998). Uncannier than my mine, his involved an old woman dictating a long poem to him from the back seat of his car:

> I took my foot off the gas and turned my head to look directly into the back seat area. As I turned, I realized that the voice was apparently turning with me, keeping its same relative distance and position, behind me and a little to my right. That could only mean that the voice was *in my head!* But without that clue, all my other perceptions told me it was a *real* voice, physically speaking into the air in the closed space of the car. (Taylor 1998: 199)[2]

This event occurred later in the year after Taylor had been writing an article on the transition from matriarchy to patriarchy. On the evening of this experience, he was tired, but highly aroused emotionally, returning home from his weekly dream group. At first fearful, he concluded excitedly that his 'muse' was giving him a poem, as Coleridge had received 'Kubla Khan'. Taylor 'stay[ed] up all night scribbling frantically, trying to keep up with the steady flow of dictation' (Taylor 1998: 200). He continued

---

1   At the Julian Jaynes conference (2014) where I presented, I was told that Steven Moore, friend of the English writer Alan Moore, heard the word 'Endymion' in a Darth Vader voice on awakening from a dream. Messages heard in an altered state can be apropos, even prophetic. Jaynes said the 'low, guttural voice' under hypnosis shows its provenance in a separate dissociative stream of speech (Jaynes 1976/1990: 392).

2   Jaynes's one voice aphorism said, 'Include the knower in the known', heard from behind and to the right.

to write 'hurriedly and continuously' on airplanes, between lectures and late at night, as his muse took off the pause button and started up again (Taylor 1998: 201).

The resultant twenty-six-page poem recalled Queen Jocasta's submission to King Laius, whose abandoned son, Oedipus, returned to kill his father and marry his mother, unaware of their true relationship. Taylor's poetic dictation, alliterative but non-rhyming, was full of Jocasta's prophecies and a longing for Mother Right. The old woman's voice fell silent after the poem ended. Taylor never channeled (his own word choice) another. While he described his experience rationally, he felt, nonetheless, that his explanation was insufficient:

> I was so interested and involved in that research that even after I was done, my unconscious continued to shape the material I was working with into more emotionally interesting and compelling narrative form. On the face of it, that certainly is what happened; but I have had many versions of the experience of 'unconscious incubation' over the years, and none of them were as complete or as 'other' as the voice of that old woman, and I reserve the right to suppose that there is more to it than that. (Taylor 1998: 202)

Both Taylor and I had 'received' wisdom from someone *totally* other, a psychotic, male figure with a Darth Vader voice and an old Greek woman, traumatized by the sight of her mother 'pitted on a stake' for telling the tale she would then unfold (Taylor 1994: 204). Our experiences encapsulated the Jaynesian phenomenon of the 'god-like' or 'daemonic' creative properties of the right hemisphere that Persinger had also seen in his 'Muse Factor' study. Both had a common link to emotional reading or writing. The voices erupted in an altered state to command or dictate what we could not consciously say ourselves.

Dissociative Voices and Conversion Experiences

A cursory look at a few biographies of religious figures and creative writers provides abundant evidence for the role of reading, writing, contemplation, meditation or even highly charged speech in provoking a dissociative feeling of presence, an alien voice or even possession states. In each of the following cases, an answer to an inner crisis found in a biblical text veered the emotional hearer towards an intensely spiritual life.

St. Augustine, weeping in the throes of a spiritual crisis, said he heard a child from a neighboring house chanting over and over again: '*Tolle. Lege* [*Pick up. Read*]'. He interpreted this voice as a divine command to pick up the book of the apostle Paul he had put down beside him. Augustine focused his attention on the first passage he saw (Rom. 13:13–14), interpreted as an order to abstain from lust. Precedents and predecessors played an important role with a contagious effect. St. Paul himself had been converted after hearing a divine command on the road to Damascus. St.

Anthony had converted after hearing a gospel imperative to sell all and give to the poor. Alypius, who was present at his friend Augustine's spiritual awakening, continued reading in the book of Romans and was converted as well (Augustine 2004).

Former nun Karen Armstrong, now a highly regarded historian of religion, described the even more dramatic conversion experience of the Prophet Muhammad. After lengthy, isolated meditation, he was 'torn from sleep and felt himself enveloped by a devastating divine presence', later described as 'an angel [that] had appeared to him and given him a curt command: "Recite! (*iqra!*)", also translated as 'Read'. Muhammad was aware of traditional Arabian 'ecstatic soothsayers', who recited oracles, as well as of poets possessed by 'personal *jinni*'. He wanted no part in this. However, after three 'terrifying embraces', the 'first words of a new scripture' poured forth (Armstrong 1993: 137). He was forced to recite against his will, often 'struggling to make sense of a vision and significance that did not always come to him in a clear, verbal form' (Armstrong 1993: 139). 'Sometimes it comes to me like the reverberations of a bell', he said, 'and that is the hardest upon me; the reverberations abate when I am aware of their message' (Bukhari, Hadith I.3, Qtd. in Armstrong 1993: 139). Both the non-verbal bell sound, known to set off seizures in epileptics, and the forced words make the case for right-hemispheric provenance.

Similarly, a terrified young French girl, Joan of Arc, heard a voice to her right, towards the village church, in the presence of great light. First she heard the voices of Sts. Catherine and Margaret. St. Michael's followed, commanding Joan to go into France from Burgundy, then a separate country under English domination, to save the French king. Joan's coming had been prophesized in local legends and her father had dreamed she would escape from home with a band of soldiers. Initially, she heard her voices 'when the church bells were sounding for prayer', but they later became clearer and more frequent (Trask 1996: 161). Joan of Arc would be dead, burned at the stake, by age nineteen, having dressed as a male to fight for French sovereignty.

The revelatory process for the Prophet Muhammad was slower and never easy: it came to him in a trance state, torn from him 'bit by bit, line by line and verse by verse over a period of twenty-three years' (Armstrong 1993: 139). Armstrong *almost* named the right hemisphere when she wrote about the commonality of all mystical experiences of God, saying: 'it is an interior journey… undertaken through the image-making part of the mind — often called the imagination — rather than through the more cerebral, logical faculty' (Armstrong 1993: 219). Muhammad's experience, she said, was akin to dissociative poetry: it is 'rather as a poet describes the process of "listening" to a poem that is gradually surfacing from the hidden recesses of his mind, declaring itself with an authority and integrity that seem mysteriously separate from him' (Armstrong 1993: 139).

Still doubtful and resisting the call, Muhammad saw a vision of the angel Gabriel as 'an overwhelming ubiquitous presence', encompassing all space around him, wherever he turned (Armstrong 1993: 138). Unable to read or write, Muhammad's first verse nonetheless referred to 'One who has taught [man] the use of the pen'. Muhammad's longing for a book for his people, similar to that of the Jews and Christians ('the People of the Book'), may have been partly responsible for highly valued revelations that must be written down, memorized and recited. His Night Journey to Jerusalem, where he met his predecessor prophets — Abraham, Moses, Jesus and others — further confirmed his mission.

Muhammad's words were so emotional, highly significant and poetic that early converts to Islam also succumbed to the 'divine invasion' while reading or listening to them (Armstrong 1993: 145). Likewise, Stevan L. Davies suggested that Jesus's followers might have undergone altered states of consciousness while listening to his highly significant, but ambiguously worded, parables. The Buddha's own intense meditative practices, still practiced by his followers, are well known for bringing out both gods and demons on the path to Enlightenment.

The great Sufi poet Rumi, a learned scholar of Islam, son of a visionary theologian and jurist, repudiated scholarly discourse, whirling his way to illumination and poetic outpourings. An alien voice had directed him to his great inspiration, Shams of Tabriz, who asked him an emotional question so shocking that Rumi fainted. The two mystics spent months together in one long ecstatic conversation. When Shams disappeared, Rumi's quest to find his Friend ended in a sense of merged presence: 'Why should I seek? I am the same as / he. His essence speaks through me. / I have been looking for myself!' (in Barks 1995/1997: xi). In other words, Rumi's ecstatic poems were coming through the exalted Friend, who would be replaced by two others as each one died.

Spurred into dissociative discourse by the intense verbal flow between him and his Friend, as well as by his whirling technique, Rumi craved the open-hearted, silent ecstasy of the unitive state after the words have stopped: 'Drink from the presence of saints / not from those other jars' ('The Many Wines' in Barks 1995/1997: 6); 'that drink is an oceanic wave hidden in the center of my chest' ('A Thirsty Fish', in Barks 1995/1997: 19). 'This is how it always is / when I finish a poem. / A great silence over-comes me, / and I wonder why I ever thought to use language' ('The Reed Flute's Song', in Barks 1995/1997: 20). 'Stop the words now. / Open the window in the center of your chest, / and let the spirits fly in and out' ('Where Everything is Music', in Barks 1995/1997: 35). 'Love opens my chest, and thought returns to its confines' ('Granite and Wineglass', in Barks 1995/1997: 104).

Many more conversion experiences can be found in William James's *Varieties of Religious Experience* and Richard Bucke's *Cosmic Consciousness*. While emphasizing the emotional instability and divided selves of mystical converts, James still considered them religious geniuses whose optimistic revelations were a worthy ideal. James distinguished those with occasional voice and vision automatisms (the Buddha, Jesus, St. Paul, St. Augustine, Luther) from the ones who seemed to be rather constantly under the 'direction of a foreign power, and serving as its mouthpiece' (James 1994: 522). Here he includes the Hebrew prophets, Ezekiel and Isaiah; the Prophet Muhammad; many minor Catholic saints; Quaker leader George Fox; and Mormon leader Joseph Smith.

Dr. Bucke had had his own epiphanic experience after an evening with two friends reading Wordsworth, Shelley, Keats, Browning and Walt Whitman. Riding home in a hansom, 'letting ideas, images and emotions flow of themselves', Bucke was suddenly enveloped in a 'flame-colored cloud', followed by a 'lightening flash' in his brain, bliss in his heart and a sense of 'exultation, of immense joyousness'. In a few euphoric seconds he became utterly convinced of the founding principle of love and the immortality of all men in a universe described as a 'living Presence' (Bucke 1961/1993: 8). Experiencing a mystical state impelled Bucke's quest to find others with similar feelings. His book portrayed the minds of religious figures, as James's did, including Moses, the Buddha, Jesus, Paul and Muhammad; but also the minds of writers: Dante, Spinoza, Blake, Balzac, Swedenborg and Whitman. Each, he said, had a 'duplex personality', with their second self objectified as a 'separate' person who embodied an evolutionarily advanced 'cosmic consciousness' (see Bucke 1961/1993: 52, 213). As to Bucke's own proclivities, it should be noted that he suffered severe anxiety attacks, a mood disorder connected to right-hemispheric over-activation (see He *et al.*, 2010).

More Famous Writers and Poets with Dissociative Creativity

Frederick Clarke Prescott, a literature professor at Cornell University, wrote a series of articles in the *Journal of Abnormal Psychology* that would later become his book *The Poetic Mind* (1922). A reviewer in 1923 pronounced it 'by far the best book of its kind yet published in America or Great Britain' (Martin 1923: 233–4). Predating Jaynes by fifty-four years, Prescott exhaustively covered the terrain of poetic creativity, present in both prose and verse, in nineteenth-century English and French writers. His examples include those who 'saw' their characters with 'poetic vision' — visual hallucinations — and those who felt split, receiving words from a second self or a hallucinatory dissociative presence. 'The death of Arthur is

poetry not because it was put into verse by Tennyson or into prose by
Malory, but because it was originally conceived by an imaginative opera-
tion of the human mind' (Prescott 1922: 8).[3]

John Bunyan 'saw figures and heard voices which were as clear and
vivid to him as those of objective reality' (Prescott 1922: 29). Charles
Dickens declared, 'every word said by his characters was distinctly heard
by him'. 'When, in the midst of this trouble and pain', he wrote, 'I sit down
to my book, some beneficent power shows it all to me, and tempts me to be
interested, and I don't invent it—really do not—*but see it*, and write it
down' (in Prescott 1922: 189, referencing Forster 1903 (3): 306-7). Visions
constantly plagued the poet Shelley, who could barely distinguish reality
from imagination. In a letter to Myers, Robert Louis Stevenson, author of
The Strange Case of Dr. Jekyll and Mr. Hyde, spoke of his 'self' versus 'the
other fellow', a contrarian side of him that made demands when he was ill
in a feverish state (Myers 1903 1: 301-2). Stevenson also claimed that 'little
people' in his dreams made suggestions for his novels and dictated many
of the details (Stevenson 1892, in Ellenberger 1970: 166). Both high fever
and dreaming produce altered states of consciousness.

Writer, poet and politician François-René de Chateaubriand conversed
with an invisible female companion. Novelist George Sand had visual
flights of imagination with a dream character, Corambé, who spilled over
into her waking life. Poet and dramatist Alfred de Musset just listened, 'as
if some unknown person were speaking in your ear'. The poet Alphonse de
Lamartine said: 'It is not I who think, but my ideas who think for me'.
Similarly, George Eliot said a 'not herself' took possession in her best
writing (in Prescott 1920: 102).

Gustave Flaubert created a character who could not distinguish dream
from waking in *La Morte Amoureuse*:

> From that night my being became in some sort double: there were two men
> in me, one of whom knew nothing of the other. Sometimes I thought myself
> a priest who dreamed each night that he was a gentleman; sometimes a
> gentleman who dreamed that he was priest. I could no longer distinguish
> dream from waking, and I could not tell where the reality began and where
> the illusion ended… Two spirals entangled in each other and mingling with-
> out ever touching will truthfully represent this bipartite life of mine. (In
> Prescott 1920: 199)

When nineteenth-century critic and historian Hippolyte Taine asked
Flaubert whether he confused his 'artistic hallucinations' with reality,
Flaubert responded 'Oui *toujours*' (in Tony James 1995: 166). Flaubert also
melded with his characters. Writing about Mme Bovary's self-poisoning,
James reported the author tasted the arsenic in his own mouth and vomited

---

3   Prescott took many of his examples directly from Frederic W.H. Myers'
    (1903) *Human Personality and its Survival of Bodily Death*.

his dinner. In this way, *'Mme Bovary, c'est moi'*, becomes more than a metaphor.

Honoré de Balzac, who consumed massive quantities of coffee and slept very little, watched his hallucinatory creations develop 'by themselves', as had Bunyan's and Dickens'. Balzac believed he could merge with people on the street, able to describe their sensations first-hand (in Tony James 1995: 160-1). Many nineteenth-century French writers also *sought out* dissociative states of consciousness through drugs and the occult, according to James.

Prescott felt that the reasoning mind plodded along, while the 'inspired moment, the crisis' poured forth from the imagination 'in one simultaneous gush' (Prescott 1922: 42). When he says that 'associative, imaginative, poetic thought is the primary one; it is older, and was indeed presumably once our only mode of thinking' (Prescott 1922: 53), he harks back to the bicameral age before Jaynes:

> The history [of primitive man], however, shows clearly first an almost exclusive use of the imagination and this giving way gradually to reasoning. It shows first a literature made up exclusively of poetry, and this gradually growing into literature made up for the most part of prose. That primitive man was more imaginative than the man of the present hardly needs proof when we remember that the earliest poetry, that of the Bible and of Homer, has always been considered the greatest, and has always been the despair of the moderns. There is reason to believe also that the dreams of primitive men were more extended and vivid than ours. The greater attention given to them, and the greater respect for them among the ancients would suggest this; and it would be expected from the analogous case of children, in whom dreaming is more vivid and absorbing than in adults... So primitive men were probably greater dreamers, as they were certainly greater poets. (Prescott 1922: 60-1)

In Freudian language appropriate for his time, Prescott said, 'in some sense and in some degree, the true poet will always be mad' (Prescott 1922: 264). But, for Prescott, 'madness' does not mean 'abnormal' or 'insane'. As he explained, 'Our old word *wood*, meaning mad, is etymologically related to *woð*, a song, and to Latin *vates*, a seer or poet, — suggesting that recognition of the poetic madness is widespread and older than Plato'. Prescott's thesis thus connected poetic madness to 'primitive peoples' and their 'priests and poets' (Prescott 1922: 265-6). He concluded, 'Poetry is therefore broadly a safeguard for the individual and for the race against mental disturbance and disease. Shakespeare, if not mad, prevents madness by writing *Hamlet'* (Prescott 1922: 274). Prescott had discovered how the act of writing itself had the power to heal, long before today's researchers (see Wapner 2008).

But Prescott spied more in the poetic mind: 'The oracles of poetry come from an unconscious visionary thought tending to transcend time and space. The poetic vision sees a future event symbolically as already present;

it sees it indeed without time; it cannot therefore establish dates according to our ordinary calendar' (Prescott 1922: 286). We will meet this concept of transcending time and space again in the next chapter.

Shelley H. Carson,[4] a lecturer at Harvard and author of many articles and several books, has produced the best research I have seen on creativity and psychopathology. She proposes a model of 'shared vulnerability', where 'cognitive systems that underlie creative ideation may be dependent primarily on irregularities in both the serotonin (governing mood, psychosis, altered states of consciousness, creativity) and the dopaminergic (cognitive processing) neurotransmitter systems' (Carson 2011: 144). A complex interplay of genes not only create the disorder, in varying degrees, but also confer traits, such as high intelligence, enhanced working memory and cognitive flexibility, that protect them. Reduced latent inhibition of incoming sense impressions, novelty seeking and a preference for complexity can foster creativity as well as overwhelm, if unchecked by the positive traits cited above.

More specifically, the genetic vulnerability, shared in bipolar, schizophrenia spectrum and substance abuse disorders (up to a point), 'may predispose certain people to experience altered mental states that provide access to—and interest in—associational material typically filtered out of conscious awareness during normal waking states' (Carson 2011: 144). Creative people also commonly report unusual perceptions and mystical experiences (in Barron 1969). Rawlings and Locarnini (2006) saw a tendency to make loose associations amongst professional artists and musicians, while biologists and mathematicians scored higher on the Asperger scale. These researchers, cited in Carson, contrast a propensity to *over*systematize in the science-math professionals with *under*systematizing in the fine arts. This finding would confirm our contentions in Chapter 1 regarding left- and right- hemispheric differences in these domains.

Abnormal hyperconnectivity, as we saw in the autistic savants, can confer super skills along with deficits. But synchronization within and between the hemispheres is a plus for creating minds. Since art, music and literary creativity are positive traits commonly conferred, Carson suggests that others on the mental illness spectrum, while not similarly gifted, might still benefit from these pursuits to help alleviate their symptoms. In summary, she emphasizes how multiple genes must interact with each other and that a person's environment 'is important in determining a

---

4   Jaynes (1976/1990) linked the Old Testament 'sons of nabiim', who remained able to hear bicameral voices, to a genetic predisposition to schizophrenia. It was actually prescribed in the Bible that they be put to death. The later prophets, Ecclesiastes and Ezra, 'seek wisdom, not a god. They study the law. They do not roam out into the wilderness "inquiring of Yahweh"' (311–12).

tendency toward either creativity or toward psychopathology' (Carson 2011: 150).

Early Childhood Trauma and Dissociative Poetic Voice

In addition to mental illnesses and techniques that produce alien voices and anomalous presences, childhood trauma is a primary factor in preparing access to dissociative creative thought. The effect of the trauma will divide along gender lines: maternal loss or an overbearing mother particularly affects boys; girls suffer more from the loss of a significant father figure or childhood sexual abuse. I would add that, for most people, we would expect the left hemisphere to produce either calming or manic presences when someone is under enormous stress. However, if a person were right dominant, we might rather expect poetic presences to arise on the right, or bilaterally.

Dr. Allan N. Schore, who is on the clinical faculty of the Department of Psychiatry and Biobehavioral Sciences at UCLA, has written extensively on the infant's relationship to the mother between the critical period of 10 to 18 months of age, and the enduring impact of early trauma on brain development. With his emphasis on the mother or principal caregiver, Schore is Freudian; but his theory is not about Freud's family romance or infant sexuality. Rather, Schore sees a *physical* alteration in the child's developing brain, which can predispose to psychological disorder.

In Schore's scheme, the mother's right hemisphere communicates directly with the right hemisphere of the infant, regulating its endocrine and nervous systems through her affirming gaze and facial expressions. Good maternal–child attachment requires synchronous rapport. In a 1 year-old child, negative emotions are already lateralized to the right hemisphere, where they remain. Growing toddlers learn to control their own emotions by evoking in memory their mother's touch and the image of her face, developmentally followed by prosodic memories of her voice. Quoting Putman's research, Schore accents the self-soothing benefit obtained by the child's 'naturally divided psychological organization as one "part" comforts another "part"' (Putnam 1992: 102). Similarly, Daniel Siegel (1999) has said a hidden observer, which helps create self-coherence, is present in both children and DID sufferers. Inner voices reflect the child's experience of 'receiving speech' from a parent or an authoritative other helping to stabilize their world (Siegel 1999: 325–6).

Reading faces and early language processing, including the child's name and single emotion-laden words like 'good', 'bad' and 'no', occur in the right hemisphere. These elementary, yet highly significant, words are crucial for developing a sense of self and must be learned in a social context, with accompanying gestures and facial expressions. Using her own right-hemispheric specialization for intonation, attention and tactile information, the mother activates the infant's right temporoparietal region (Schore, in Baradon 2010). Whether loss of the mother or an overbearing, abusive or neglectful one, the resultant damage to the prefrontal cortex and limbic areas of the right hemisphere, where the implicit, body-based self resides, will predispose the child to dissociation when stressed (Schore, in Baradon 2010). Failures in early attachment are associated with high dissociation in adult psychiatric patients. Right-hemispheric activation increases during highly emotional psychotherapy sessions (Schore 1994).

All of the above may explain why the early developing right hemisphere remains responsible for the emotional and prosodic features of language; as well as, ultimately, for the concrete, bodily- and environmentally-based language of poetry. Left-hemispheric development burgeons around age four to six, as it processes the lexical, semantic and syntactic elements of language for most. In view of these linguistic and hemispheric differences, one can imagine how the overstressed adult mind could regress to an earlier hemispheric organization, a kind of return to the internal right-hemispheric mother, where the hallucinatory calling of one's name, commands or poetic dictation redolent of rhythmical motherese might palliate anxiety and stabilize the self. Early childhood dissociation, a defensive maneuver of the autonomic nervous system, may presage the adult capacity for self-organizing imagery, dissociative poetry or consoling words, as though from an authoritative 'Other'.

Schizophrenia develops in identical twins with high frequency, up to 50%. However, in a study of identical twins in which only one had developed schizophrenia, the afflicted twins showed anatomical differences in brain volume (Goleman 1990). Using fMRI scanning, the researchers found larger ventricles filled with cerebrospinal fluid taking up space that would have otherwise been dedicated to cortical tissue. Also, wider spacing in the folds of the cortex suggested brain atrophy or a developmental failure. The left temporal lobe was reduced in size as well.[5] As with children with left-hemispheric epilepsy, we might also suggest that this reduction in the left temporal lobe might lead the right to compensate and become dominant for language.

But why did only one identical twin develop the disease? Schizophrenics are emotionally vulnerable in stressful environments, including bullying and sexual abuse (Kuipers et al. 2006). Childhood sexual or physical abuse and neglect can lead to auditory hallucinations and delusions. Trauma can change the brain physically to resemble the brains of schizophrenics. Both traumatized children and schizophrenics show hippo-

---

[5]   Shenton et al. (1992) correlated schizophrenic thought disorder and auditory hallucinations with tissue loss in the left temporal lobe causing inappropriate 'linkages'.

campal damage, cerebral atrophy, ventricular enlargement and reversed cerebral asymmetry, where the right hemisphere is larger. Tellingly, twins predisposed to schizophrenia were more likely to develop the disorder if they had been adopted into a dysfunctional family (Read *et al.* 2005). Maternal sensitivity can suppress negative gene expression, while *insensi*-tivity can express it (Popper *et al.* 2008). In a study of monkeys, a gene for aggressivity was only expressed in those with insecure attachment to their mothers during infancy (Suomi 2005). Negative parenting practices are especially potent triggers for chemical changes that influence a person's mental stability (Higgins 2008).

Jamison (1993/1994) found a high prevalence of bipolar disorder in literary and artistic families, indicating a genetic origin in this pathology. However, more recent studies show that not all children with a genetic predisposition manifest this illness. Again, early childhood relationships can determine whether or not it will manifest from a predisposition. Similarly, maternal depression dampens left-hemispheric functioning, with relative increases in right-hemispheric activation in both mothers and children. Freudian Juliet Mitchell believed that 'replacement' by a younger sibling could provoke 'hysterical' symptoms in the older child.

Traumas can have a cumulative impact as well. In a study of children ages one to five exposed to the 9/11 disaster in New York, those who had suffered a prior trauma were 'about 20 times as likely to show signs of depression, anxiety, or attention deficits as children who had not known a significant trauma before Sept. 11' (Carey 2008). The fear-generating amyg-dala overreacts chronically once traumatized. After experiencing sequential traumas, including the threat of Mongol invasion, flight and the loss of his mother in early childhood and later his great Friends, the Sufi poet Rumi intoned: 'Don't turn your head. Keep looking / at the bandaged place. That's where / the light enters you' ('Childhood Friends', in Bark 142). Similarly, English poet Ted Hughes believed all art originated with a wound. Here, he eloquently describes Shakespeare's dramatic art as a repetitive shamanic ritual:

> The secret of Shakespeare's unique development lies in this ability (in most departments of life it would be regarded as a debility) to embrace the inchoate, as-if-supernatural actuality, and be overwhelmed by it, be dis-mantled and even shattered by it, without closing his eyes, and then to glue himself back together, with a new, greater understanding of the abyss, all within the confines of a drama, and to do this once every seven months, year after year for twenty-four years. (Hughes 1994: 479)

Mother Loss and Other Trauma in Exceptional Writers

A brief look at the childhoods of some exceptional writers reveals their early traumas, along with high intelligence and an obsession with the written word. John Bunyan, raised in poverty, lost his mother at sixteen. When his father remarried within two months, Bunyan left home to join the military. A scrupulous child, Bunyan had been obsessed with his 'sins' and heard a heavenly voice telling him to stop sinning or he would go to hell. This early hyperreligiosity impelled his 30-year career of preaching and writing. Even while imprisoned, he continued to preach and published his confessional autobiography, *Grace Abounding to the Chief of Sinners*, during a brief period of freedom.

The second of eight children, Charles Dickens was relegated to a nurse shortly after birth. At 12 years old, he was forced to work in a shoe polish factory while his father went to debtors' prison. When he was sent away from this heinous workplace, his mother tried to send him back, while his father insisted he return to school. Dickens remained resentful of her intervention and ever sympathetic to the plight of children, as evidenced in his many books.

Robert Louis Stevenson exemplifies the effect of overbearing parenting and physical illness. He was the happy, but sickly, only child of parents who adored him, as well as the sole charge of a loving nurse. They read to him constantly as a small child and he dictated his own stories and letters to his parents before he learned to write. His mother would become hysterical if he suggested traveling alone. His strict, melancholic father fought with him for rejecting religion as a university student. Because of his sickliness, Stevenson was often confined to bed, becoming an obsessive reader, incessant letter writer and prolific author. Bunyan, Dickens and Stevenson appear on Jamison's Appendix B naming writers with bipolar disorder.

Jean-Jacques Rousseau lost his mother to puerperal fever three days after his birth. He was made to 'grieve for a mother whom he resembled disturbingly and had somehow killed' (Damrosch 2005: 7). After this initial loss, Rousseau's beloved aunt, Suzon, as well as his unstable father, would abandon him as well. Rousseau was shy, agoraphobic, insecure, melancholic, a hypochondriac and, eventually, paranoiac (Damrosch 2005). His sexuality permanently compromised, he sought consolation in older maternal figures, such as Mme de Warens, whom he called '*Maman*'. These twinned souls had both lost their mothers; but she had also lost a mother figure and her father, all before she was 12. Rousseau's love of nature could be construed as a return to the Mother. Fortunately for future French children, he insisted that mothers, not wet nurses, care for their children, thus revolutionizing child-rearing practices, despite his own disavowal of the children he fathered, who were relegated to an orphanage.

Both a philosopher and a novelist, Rousseau found sublime pleasure in music and the winged flights of his imagination. Rejecting the rational for the intuitive, he experienced an epiphanic moment traveling on the road to

Vincennes on his way to visit his imprisoned friend, the philosopher Diderot. In one bright flash of understanding, Rousseau recognized the inherent goodness of all men, in an otherwise corrupt society, and refuted the pretensions of all major revealed religions. His first *Discourse*, conceived in a divine light yet unerringly his own, claimed all men could find a divine voice in their own conscience. His predilection for solitude bred more mystical oneness. He claimed to have composed the entire fifth chapter of *Emile*, his great treatise on education, in a continual state of ecstasy. Yet he still felt the need to justify himself to his detractors, writing a book where he literally split in two—*Rousseau juge de Jean-Jacques* (*Rousseau Judge of Jean-Jacques*)—as well as his exquisite *Confessions*.

French novelist Honoré de Balzac was born one year after his parents' first child had died. Rousseau's admonitions notwithstanding, Balzac was delivered to a wet nurse a few hours after birth, not returning home to his parents until he was 4 years old. His situation not much improved, as he was subjected to the disapproving ministrations of his mother and a governess. He was sent to day school before turning 5, then to an out-of-town *collège* (middle school) when 8. He saw his mother only twice during the next six years. Refusing to do any work at school, he was sent to an alcove under the stairs four times a week as punishment. Here, he 're-created in his mind's eye scenes he had read about in books, observing his own thoughts' (Robb 1995: 16).[6] Balzac's eidetic memory served him well. He was 'an exuberant mind raised in solitary confinement', who read most of the books in his school library (Robb 1995: 18).

Balzac's mother produced another son, fathered by a local landowner, whom she obviously preferred and lavished with love. Balzac spent his life searching for maternal replacements. The deprivations of his early years spawned a genius with incomparable mental powers for literary creativity, along with mystical and androgynous leanings.

Guy de Maupassant, another prolific French writer of the nineteenth century, suffered his parents' separation at 10 years old. His mother's nervous problems interfered with their relationship and she was fundamentally cold towards him. Although she had always encouraged his writing, she remained distant, not visiting him in the asylum where he had been committed after a suicide attempt. He died institutionalized and alone at 43, his brilliant mind destroyed. Affect trauma, along with a genetic predisposition to mental illness (his brother succumbed first and died before him), plus tertiary syphilis in an age without penicillin, all played a role in his early demise.

---

6   These and other details on Balzac's life are from Robb's (1995) biography of the writer.

Nonetheless, or most likely as a saving grace, Maupassant wrote continually, producing three hundred short stories and six novels in eleven years. Hardly a jumbled, nonsensical hypergraphic, he sculpted his sentences into a triune pattern of adjectives and subordinate clauses, along with careful overall symmetry. The tight coherence suggests a nearly compulsive left-hemispheric style, helping to tame his disordered emotions. In *Pierre et Jean*, the ending revisits in exact opposite order the expositional scenes from the opening chapter. Yet, Maupassant's tortured inner world intruded into his writing, with a hatred of women, maternal betrayal and anomalous occurrences mirroring his own, including a hostile sensed presence in 'Le Horla'. In 'Madame Hermet', a sick boy, whose mother refuses to visit him for fear of catching his illness and ruining her beauty, dies alone. As in the story, so in life: Maupassant never saw his mother again, despite repeated requests.

The overpowering impulse to write can be an emotional reaction to troubling internal or external events. *Associative* thoughts and images brewed from the repertoire of the mind/body connection are *normal* responses to strong emotions (see Damasio 2003). However, *dissociative* thoughts and images are the mind's reaction to unbearable emotional events, defenses deployed to stabilize the disintegrating self. Van Gogh is a notable example of a creative artist who both wrote *and* painted compulsively. 'At his peak, he painted a new canvas every thirty-six hours' and wrote 'two or three six-page letters a day' to his brother, Theo, according to Dr. Alice Flaherty, who became hypergraphic in the wake of her twin babies' death. Echoing Blake, he said, 'Sometimes I draw sketches almost against my will... The emotions are sometimes so strong that one works without knowing one works, when sometimes the strokes come with a continuity and coherence like words in a speech or a letter' (Flaherty 2004: 74–5).

Flaherty attributed van Gogh's compulsive production to epilepsy and bipolar disorder. He also appears to have been emotionally unstable from early childhood, a stranger with eight brothers and sisters in a cold, gloomy, sterile atmosphere. He was born exactly one year after his mother's first stillborn child, named Vincent before him. With a persistent sense of failure and unworthiness, he once commented that his brother Theo was the one 'who comforts his mother and is worthy to be comforted by his mother' (see Butterfield). His emotional instability may have been triggered by a grieving mother unable to create a strong enough maternal bond, leaving him permanently volatile, especially when subsequent losses threatened his well-being. His mental breakdowns often occurred when he perceived a threat to his relationship with loved ones, such as Theo, or his artistic companion, Gauguin.

Presence in Women Poets and Mediums

Martin Teichler, an associate professor of psychiatry at Harvard Medical School, has discovered that boys are more sensitive to neglect and girls are more vulnerable to sexual abuse, with both genders showing a smaller corpus callosum, hence less connectivity between the hemispheres. Patients with dissociative identity disorder, usually resulting from childhood abuse, report more paranormal experiences than patients with any other type of disorder. Voices are constantly present; patients enter trance states instantly and have spontaneous past life intrusions; conflict about sexuality and sexual orientation is nearly universal. While many studies have said that women are less lateralized for language than men, meaning language functions are more spread out in both hemispheres, others do not. But, if Schore is right, the mother's right hemisphere should be more potent, synchronizing with her baby in ways a father's cannot. In addition, girl children are more apt to suffer sexual abuse, making it more likely for them to experience dissociative voices. The fact that the oracles were always women, working at the behest of male priests in ancient Greece, shows an early tradition of female dissociation.

While women have more of a tendency to dissociate, what they have historically lacked is respect, a voice of authority and their own literary and/or mystical tradition. Rather than create a harmonizing presence, dissociating women poets tend to break down in this lack, criticizing both their worth and their femininity — *unless* they have a male collaborator. Trapped in societal gender expectations or the very real difficulties of combining motherhood and creative projects, their voices and visions may become tormentors, not protectors (Weissman 1993). Without a defensive narrative web and social support (from this world or beyond), women may succumb to suicide, rather than seek assurances of immortality. Virginia Woolf, who was abused by her half-brothers and lost a neglectful mother as well as a brother, said that '[her] pen seemed to stumble after [her] own voice, or, almost after some sort of speaker, as when [she] was mad' (Qtd. in Claridge 1990: 190). Anne Sexton, another abuse victim, had full-fledged dissociative identity disorder. She claimed *she* was not a poet; her poetic alter did the writing; she just sold it (Ross 1994). Both women committed suicide.

The loss of an important male figure is also significant for women. Sylvia Plath's poetic genius had sprung from the loss of her father whom she sought in her Ouija board sessions with her husband and in her poetry. However, when Hughes abandoned her, she ended her life. Emily Dickinson began writing poetry after a beloved minister friend moved away, but continued to produce highly unusual, imagistic verse, bolstered by a literary critic with whom she corresponded in her remaining years of life.

In the mystical world, the Seeress of Prevorst was sent to live with her grandparents at 5 years old. Here, she was most likely sexually abused. She fell ill after the death of a minister friend, with whom she had conversed regularly, and a forced marriage to another man. Subsequent to a dream in which the corpse of the minister lay next to her in bed 'healing' her, nightly trance states began with paranormal prescriptions for healing 'based upon the magical power of words and numbers'. Coming from a folk tradition steeped in magic and the occult, her reaction to abuse and loss could be foreseen, but a doctor who took her into his home and encouraged her dissociative outpourings only intensified it (Hanegraaff 2001: 217–18).

Literary scholar Dianne Hunter adds the tale of 'Anno O.', Freud's famous German patient, Bertha Pappenheim. She was born to a German Orthodox Jewish family with an authoritarian mother, after two sisters had already died in childhood. With relatives who had been psychotic and in the sole care of her ailing father, Pappenheim not surprising exhibited 'hysterical' symptoms: paralysis on the right side of her body; writing only with her left hand; 'intermittent deafness'; disorganized speech; and finally aphasia, despite her intelligence and talent in languages and poetry. She regained her speech, but only in foreign languages, and suffered visual hallucinations. 'Pappenheim claimed to be divided between two selves, "a real one and an evil one which forced her to behave badly", as well as a hidden observer. Two states of consciousness would alternate, one of which would interrupt while the other was speaking' (Garner *et al.* 1985: 92). Under hypnosis, she spoke only in English. Her dissociative behavior culminated in a hysterical childbirth, forcing her doctor to abandon her case.

Hunter's comprehension of the early connection to the mother is profound: 'Before we enter the grammatical order of language, we exist in a dyadic, semiotic world of pure sound and body rhythms, oceanically one with our nurturer' (in Garner *et al.* 1985: 89–99). Pappenheim's dissociative personas effectively cured her through body language, leading to a full recovery with a mission: she became a feminist, headmistress of an orphanage and entered a career in social work, 'rescuing and sheltering abandoned and abused women and children', including Jewish girls who had been sent to work in houses of prostitution under false pretenses (in Garner *et al.* 1985: 103).

As we will see, the male poets Hugo, Rilke, Yeats and Merrill primarily sought the advice of dead poets and prophets through their spiritualist techniques. Spirit mediums of the nineteenth and twentieth centuries were mostly women transmitting the advice and texts of dead male poets and novelists who spoke through them (Sword 2002). The most famous channelers of the New Age movement were also women whose possessors were highly verbal males, such as Jane Roberts' Seth and Helen Schucman's voice of Jesus.

***

Persinger's "Muse Factor" experiment helps us understand how a dissociative sense of presence can be created in mystics and poets with strong emotional reactions to verbal meaning. Along with a genetic predisposition to certain mental disorders, we have proposed the additional factor of adverse childhood circumstances, especially maternal deprivation or smothering in boys and sexual abuse or loss of a father figure in girls. Other emotional stressors later in life can further sensitize the self to creative writing. The wounded mind searches for enhanced personal meaning and stabilization through a restorative dialog with conjured Others and real collaborators. The right hemisphere can facilitate the quest by commanding action, taking over motor control, dictating words, suggesting further readings or possessing the vocal chords to speak. In order to have output, there must be input. High intelligence along with voracious reading can provide the materials, the impulse and the control to pound personal emotion into eternal art.

Chapter Four

The Medium and the Matrix: Unconscious Information and the Therapeutic Dyad

Spiritualism, Jung and Flournoy

In 1848, two young girls, the Fox Sisters, claimed to get messages rapped out by spirits on the walls of their Hydesville, New York, farmhouse. They interpreted the messages using a knocking alphabet. This sensationalist, highly publicized experience jumpstarted the spiritualist craze, hitting Europe like a psychic storm in the wake of their tumultuous political revolutions of the same year. A direct heir to these influences, Victor Hugo spent two years in table-tapping séances with family and friends, receiving revelations from loquacious dead contacts, as literate as their esteemed 'transcriber'. Helena Blavatsky, a Russian import to America in 1873, who claimed Tibetan Masters for contacts, also played a large role in the spiritualist phenomenon. In 1875, she formed the Theosophical Society with worldwide branches and converts, including another great poet, William Butler Yeats. Yeats also attended séances and channeled 'communicators' through his wife's automatisms.

Many important psychologists frequented mediums, including Freud, Jung, Bleuler, James, Myers, Janet, Bergson, Richet and Flournoy. Early in his career, Jung downplayed the association between trauma and dissociation, despite his mentor Freud's early (1895), later repudiated (1905), connection between childhood sexual trauma and hysteria. In Jung's doctoral dissertation (1902/1977), he recounted the spiritualist experiences of his own cousin, Helene Preiswerk, dubbed 'S.W.'. As background, he mentioned her large family with fourteen siblings, a 'brutal' mother and a frequently absent, then recently deceased, father. He emphasized her 'poor inheritance', i.e. family members, going back to her grandfather, who were eccentrics and hysterics having clairvoyant visions and 'uttering prophesies', as well as her 'mediocre education' and preference for 'daydreaming' (Jung 1902/1977: 20–1). In all likelihood, both genetics and a harsh childhood environment played a role in her dissociative tendencies.

Jung did not treat his cousin, who suffered from crushing headaches and would become possessed by her 'spirit' relatives both in séances and out on the street, along with OBEs. Rather, he engaged her spirits out of scientific interest and curiosity about the occult, in private séances and others where his mother and other relatives were present. Skea (2006) suggested that Jung might also have been attempting to deal with his own dissociative splits by studying his cousin's experience. Jung further encouraged his cousin, giving her a copy of Dr. Julius Kerner's book about the Seeress of Prevorst (1845). Kerner, both physician and poet, believed in and had encouraged his patient's 'spirit' contacts.

Elements of Kerner's book found their way into Helene's trance states with reincarnation themes. At one point, Helene, in her trance role as 'Ivenes', professed to have *been* this famous medium in a former life. In fact, she invents a complex array of interrelationships where in the eighth century she was the mother of her current father, as well as of her grandfather and her cousin. In the thirteenth century, she was Jung's mother and burned as a witch. Because Helene was a young teenager when these séances were taking place, Jung attributed her reincarnation fantasies to the sexual wish fulfillment and the fertility impulses of puberty. Sex and violence did predominate, as each 'family romance' had a 'gruesome character' with murders, seduction and banishment (Jung 1902/1977: 40–2).

Influenced by Pierre Janet (1889), Jung considered his cousin's trance personalities dissociative personality fragments or 'complexes'. Cryptomnesia, the retrieval of forgotten memories of things fleetingly read, seen or heard in the past, was Jung's possible explanation for Helene's 'heightened performance' in trance, 'her knowledge of high German and customs from earlier times, despite her limited education, and facts about long-dead ancestors' (Skea 1995/2006). Jung admitted that the mystical system she constructed in trance state was 'something quite out of the ordinary' for someone so young, while noting the 'parallels with our gnostic system, dating from different centuries, but scattered about in all kinds of works, most of them quite inaccessible to the patient' (Jung 1902/1977: 91).

Théodore Flournoy, a professor of philosophy and psychology, studied the medium 'Hélène Smith' (actually Elise Müller, who worked in a silk shop) from 1894 to 1901, around the same time as Jung's séances. Hélène Smith also constructed reincarnation fantasies around her interested professor, expressed through audio and visual hallucinations, as well as in table-tapping and possession by her 'spirit guide', Léopold. The 'spirit' of Victor Hugo had previously written poetry to Hélène, while serving as her principal guide, before Leopold deposed him.[1] In two previous existences, Hélène claimed she had been Flournoy's mother; in a medieval Hindu existence, Flournoy was her beloved husband with whom she was doomed to die on a funeral pyre. The need to both 'be' a mother and please the father/lover seems apparent in both 'Helenes'.

Initially tantalized by the possibility of a former life, Flournoy searched for evidence linked to names, places and dates his subject had proffered in trance states. In the end, he rejected any 'supernormal' explanation for corroborating facts found in dusty old books or town archive records, proposing the less likely, and firmly denied, possibility that she had seen and forgotten them. He suspected that she had also picked up information from school, relatives and/or conversations and 'unconscious muscular contractions' of sitters influencing the table's revelations. For instance, the connection between Léopold and Cagliostro seemed to have been implanted at the suggestion of a woman hosting the séances where he began to appear. The 'spirit' first introduced himself as Léopold, but later said he was a reincarnation of Cagliostro, without changing his name. Interestingly, Yeats's spirit guide would also be called 'Léopold'.

Since Flournoy was more interested in linguistics than romances, his compliant subject constructed both a 'Sanskritoid' language, confirmed as such by the eminent linguist Ferdinand de Saussure who attended a session, and a 'Martian' language for his benefit. Both Daniel Rosenberg's (2000/1) article about the collaborative aspect of Hélène's linguistic pro-ductions and Mireille Cifali's (1994) appendix to Flournoy's (1901/1994) book are very informative. Although Hélène's father was a polyglot, she professed to be unable to learn foreign languages. Yet, she was able to construct her own languages in an altered state, prodded by an interested professor, whom she fantasized as both son and lover.

But how did she generate the words in her made-up languages? A possible answer lies in Newberg's (2007) fMRI research on Pentecostal glossolaliacs, whose streams of sounds are not actually words, but purport to express deep felt meaning. Newberg found significant increased activity in the temporal lobes and cited Persinger's research. Newberg also found that when frontal lobe activity decreased, the religious practitioners described a sense of giving up their will to 'the presence of God'. While their parietal lobes increased slightly in activity, suggesting they had retained their sense of self, the *language areas did not change at all*. Without

---

1   Hélène Smith must have known about Hugo's séances, so his tenure as her first guide is not surprising. Later, when André Breton, leader of the Surrealist movement, consulted a clairvoyant in Paris, she told him he would be preoccupied with a woman named 'Hélène. Shortly thereafter, indeed he was. Breton famously chronicled his relationship with a mad woman, Nadja, who announced, '*Hélène, c'est moi*'.

frontal lobe control, their words were 'loosely strung together' and merely repeated 'familiar phonetic sounds' (Newberg 2007: 195). There may be correlates between the unwilled speech of glossolalia and other linguistic automatisms like spirit possession, mediumship, table-tapping, automatic writing, Ouija board messages and New Age channeling.

While Hélène Smith may have tried to gain Flournoy's heart by speaking in tongues, her portrayal in his book angered her to the point that she barred him from attending further séances. Rather than heal his subject, Flournoy had cultivated her dissociative tendencies in the interest of science and for his own benefit, much like Jung had done with his young cousin. Jung's Helene died from tuberculosis at the age of 26, reverting to the mentality of a 2 year-old before falling into 'her last sleep' (Jung 1961: 107). Flournoy's Hélène received money from an American spiritualist that allowed her to quit séances and become a trance painter of religious imagery. After a dear Italian friend died, however, Hélène abandoned painting to seek spiritual contact with the lost friend. For both 'Helenes', spiritualism seems to have been a way to make loving connections under the veil of the trance state.

Jung and Flournoy were both drawn to occult phenomena while officially denying their existence. Both recognized their subjects 'mediumistic', i.e. dissociative, abilities. Flournoy (1902) claimed the main difference between mediums and ordinary people 'is that with the latter there is practically a very marked trench between dream and waking... With mediums on the contrary... there is no stable barrier between sleep and waking' (Flournoy 1902: 127, trans. S. Samdasani). At the time, neither theorist suggested their subjects could be receiving unconscious information directly from them, the specialists. Both, through their interest, relentless questioning and repeated sessions, pushed their subjects on to further splits.

Jung, due to his own uncanny dissociative experiences and those of schizophrenics he treated at the Burghölzli mental hospital, would go on to posit the existence of the collective unconscious, tantamount to the ancestral memories of the whole human race, accessible to the personal unconscious and from there to consciousness, to explain otherwise unknowable information expressed while in trance. He also credited the possibility of a 'transpsychic reality underlying the psyche' that touches 'on the realm of nuclear physics and the conception of the time-space continuum' (Jung 1902/1977: 125 fn. 15), further stating that 'telepathic phenomena are undeniable facts' (135).

'As if' Communication with Dissociative 'Entities'

Medical as well as lay interest in spirit mediumship was a particularly nineteenth-century phenomenon, where 'contacts' with the dead were actively sought and studied. Yet, twentieth- and twenty-first-century doctors have also treated their patients' dissociative personalities 'as if' real. Dr. Colin Ross, a Canadian specialist in the field of dissociative identity disorder (DID), treated a young woman who had been placed in the care of her grandmother after her parents had died in a car accident, yet had continued to hallucinate her grandmother's negative comments after she passed away. Using a marriage counseling-style therapy to reconcile the woman with her deceased grandmother, Ross conducted a prayer ceremony, with a chaplain present, effectively dispatching the offending relative off to 'heaven'.

Of course, sanctioning the existence of one 'entity' may lower the threshold, allowing others to emerge and multiply. Therapists can bring forth memories for events that never occurred through suggestion, which could explain the incremental growth of 'recovered' memories of sexual abuse. The multiplier effect seems apparent in Jung's cousin's dauntingly complex relationships, as she identified her reincarnated spirit in myriad ancestors. However, one telling séance session may have uncovered sexual abuse in her case as well (see Skea 2006). Ross firmly believes that 'virtually all symptoms in psychiatry are potentially trauma-driven and dissociative in nature' and that 'severe, chronic childhood trauma is a common trigger' (Ross 1994: xiii; 70).

Neurological Evidence for Dissociative Identities and the Role of Childhood Trauma

Philosopher Ian Hacking (1995) presented the contrary case that dissociation is more like storytelling and child abuse is often a story brought forth by therapists looking to cure their patients through the startling discovery. However, a more recent theorist, psychoanalyst and traumatologist Elizabeth Howell (2005) strongly disagrees, citing 'shockingly high' rates of child abuse (Howell 2005: 16) with corroborating evidence from Putnam (1997), Siegel (1999), Gold (2000), Lyons-Ruth (2003) and Vermetten *et al.* (2006). What Freud called a 'neurosis' can now be considered 'post-traumatic stress and dissociation' (Howell 2005: ix). The attachment theorists above, along with Schore, believe that not just severe abuse, but also parental neglect, deprivation and rejection in childhood can predict future dissociation.

Additionally, what Freud called the 'unconscious' could actually be 'different loci of unconscious memory that are stored in the body memory' (Howell 2005: 28). Michael Jawer and Marc Micozzi (2009) believe they have confirmed this insight. Since the fear-processing amygdala is already operating at birth, maturing faster than other brain structures, infants and young children will tend to hold on to their traumas, even if they cannot consciously recall them. Like Schore, Jawer and Micozzi believe the problem lies in the orbitofrontal cortex. A 'dampening' of function in this part of the neocortex results in a lowered 'set-point' for their hypothalamic-pituitary-adrenal system (HPA), making young trauma victims hyper-vigilant, i.e. on constant alert for new dangers. Even maternal stress while pregnant can contribute to heightened sensitivity in the child. Negative energy frozen inside the nervous system is unable to be discharged.

Jawer and Micozzi believe that the trauma is predominantly held in the right hemisphere and cannot be excised except through body-based therapy, such as massage and yoga. For this reason, we often find that those who experience anomalous perceptions, such as ghosts, poltergeists, mysterious raps on walls and moving objects, are those who had previous childhood trauma or suffered a great loss. The spiralling vortex of repressed feeling trapped in the body explodes as electromagnetic energy in the immediate surroundings, producing bizarre sights and sounds in some people, rage or extreme panic in others. Rational, impulse-controlling consciousness cannot contain the uprush of negative emotion or easily interpret the event in words.

We might also add that mediums, a special subset of dissociatives, could create authoritative, knowledgeable alter personalities able to escape the clutch of body-based fear and use the words of these protector selves to console, stabilize and even heal. Jawer and Micozzi go so far as to say that persons who experience strong emotions during life crises could *project* imagery into the minds of their close connections. A dying person and the dreamer who 'sees' him or her at the foot of her bed is an example. This type of projection might also be compared to a therapist passing on information, unknowingly or consciously, to a patient's mind, in an effort to instill a healthier version of negative events and to affect a cure.

Accepting the genesis of dissociation in severe childhood trauma, Berlin and Koch (2009) cogently argued the case for distinct neural networks in different dissociative identity states, citing two important studies. In Waldvogel *et al.* (2007), a 33 year-old dissociated woman, with no organic injury to the eyes, had become subjectively blind after a reported head trauma. In psychotherapy, she presented with ten personalities possessing different names, ages, genders, attitudes and proclivities. Differences in temperament, voice and gesture were also distinguishable as well as in languages spoken: English (she had spent a few years of her youth in an English-speaking country), German or both.

After four years of psychotherapy, the patient, in one of her younger male identities, was suddenly able to read a few words from a newspaper headline immediately after a session. This capability evolved into total vision. While more personalities came forth in the course of treatment, fewer of them were totally blind, and different visual capabilities could alternate in seconds. EEG tests performed on both a seeing and a 'blind' personality, evoked by calling his name, confirmed that the former had normal vision, while the latter had almost no visually evoked activity in response to the same stimulus. This implies that in psychogenic blindness, 'the brain can rapidly intervene at a very early stage of the visual system, preventing visual information from reaching the patient's cortex' (Berlin and Koch 2009: 19).

In the second study (Reinders *et al.* 2006), eleven DID patients were given PET scans to contrast their physiological and cerebral blood flow response to a traumatic versus a neutral autobiographic manuscript read by a psychiatrist on audiotape. The researchers found that only the traumatic identity state remembered and emotionally responded to the traumatic manuscript, while the neutral identity state had no such response and claimed no recall of the event. Berlin and Koch hypothesized that dissociation may result from 'the failure of coordination or integration of the distributed neural circuitry that represents subjective self-awareness' (Berlin and Koch 2009: 19).

The Role of the Right Hemisphere in Dissociation

As we saw, Allan Schore (2009) emphasized the need for a secure attachment to the mother for an infant to maintain its internal homeostatic equilibrium. The child's core self develops via interpersonal relations communicated through the emotion-processing right hemisphere in the first year of life. Schore specifically located the brain's major self-regulatory systems in the orbital prefrontal areas of the right hemisphere. Devinsky (2000) said that 'the essential function of the right lateralized system is to maintain a coherent, continuous, and unified sense of self' (in Schore 2009: 195). Schore also cited Molnar-Szakacs *et al.* (2005) who further summarized: 'Studies have demonstrated a special contribution of the right hemisphere (RH) in self-related cognition, own-body perception, self-awareness, autobiographical memory and theory of mind' (in Schore 2009: 195). Stressful, negative parenting can disrupt that continuity as the child defensively switches from autonomic hyperarousal to an energy-conserving dissociative state. Later stressors will also disrupt the usually integrated functions of consciousness. Schore cites two other studies that show a predominant role of the right hemisphere in representation of the self-concept (Reinders *et al.* 2003) and in dissociating psychiatric patients (Lanius *et al.* 2005).

Writer, teacher and former dissociative, Elizabeth Mikal, first got in touch with her alter personalities, for whom she had been amnestic, when a therapist suggested she write a diary using her non-dominant left hand. In the book she later published, Mikal's (2005) description of her breakdown sounded like a combination of epilepsy and psychosis, but was actually a severe case of dissociative identity disorder commensurate with the sexual and physical abuse she had suffered as a child, with right-hemispheric personalities overwhelming the left hemisphere:

> With this surrender came a collapse of my physical body. I was admitted to the hospital in June 1992 with unexplained seizures. For three days my right side felt paralyzed. I could hardly speak, and the seizures came fast and furious. At times I was aware of being in the hospital, but mostly I was aware of a sense of invasion. It felt like I was possessed. I heard voices and felt movement that appeared disconnected to my body. When I could speak, I heard myself articulate in several voices. My movements took on a variety of expressions. It was not until I was home recuperating that I felt safe enough to let the personalities emerge one by one. (Mikal 1995: 11)[2]

Going back to the nineteenth century, signs that the dissociated self-states involved the right hemisphere were descriptively evident in Flournoy's research. For example, while Hélène pantomimed in trance her reincarnation fantasy, her spirit guide, Léopold, explained what was happening, telling the doctor what to do by tapping on the table with the index finger of her *left* hand. He told Flournoy to press his thumb on her *left* eyebrow when he wanted her to wake up. He usually spoke into Hélène's *left* ear, from a distance of six feet or more. He also wrote in a completely different handwriting when using her hand and dictating poetry. At times, Hélène unconsciously switched to Léopold, speaking in a deep male voice then back to her own, without acknowledging the switch. Flournoy only remarked that Hélène suffered from 'allochiria', a condition confusing the left with the right side of the body, when entering trance states.[3] However, it seems as though she was confabulating the reincarnation stories in her left hemisphere,[4] while her male alter spoke in an accent, recited poetry and made commands under the right's control.

---

[2] Mikal's description helps us understand 'Anna O.' who also suffered right-sided paralysis and a confusion of voices speaking in foreign languages.

[3] Léopold first appeared to Hélène as a priest figure when she was 10 years old, chasing away a large dog that had frightened her. He appeared again, barring her from entering certain streets on her way home to protect her. Léopold also claimed to be Giuseppe Balsamo, a.k.a Cagliostro, a clairvoyant Italian adventurer and lover of Marie Antoinette. Reincarnated in Hélène, Marie Antoinette is supposedly reunited with Cagliostro through Léopold.

[4] See Gazzaniga (1998) on the left hemisphere's 'interpreter' function, which 'constructs theories to assimilate perceived information into a comprehensible whole. In so doing, however, the elaborative processing has a deleterious effect on the accuracy of reconstructing the past' (Gazzaniga 1998: 26).

Dissociation and Telepathy

In the nineteenth century, Frederic Myers, an English poet and classical scholar, undertook research on the survival of death in a large two-volume work. He referred to dextro-cerebrality (right-hemisphere dominance) and sinistro-cerebrality (left-hemisphere dominance) respectively, crediting the former with a greater capacity for telepathy. He detailed numerous examples of telepathic communication from the 'disincarnate' to the living, either written, spoken or appearing as a visual image, during moments of crisis or at the hour of death; in other words, in highly stressful conditions. He firmly believed that 'the human spirit' could have 'direct knowledge of facts of the universe' outside of normal sensory awareness (Myers 1903 1: 11).

If dissociative identities do indeed arise out of a disordered sense of self in the right hemisphere and the right hemisphere has some connection to non-local awareness, Ross's allusion to the many paranormal claims he heard in his patients, including 'telepathy, telekinesis, clairvoyance, seeing ghosts, poltergeist contacts' and past life intrusions, makes sense, whether real or imagined (Ross 1989: 108). He stated that his patients exhibited a 'sense of higher intelligence' in the dissociative state, due to the 'complexity of the system and the vast quantity of information it organizes, stores, and accesses' (Ross 1989: 119). This may partly explain their complexity, but the altered dissociative state itself, with its broader access to unconscious materials, may explain the high-systematizing capability, similar to what we have seen in autistic savants who had hyper-connectivity in some brain areas and shutdowns in others.

Alternative healers, whose abilities are often preceded by trauma, come to rely on *volitional* dissociative experiences (channeling, spirit guides, telepathy, distance healing), not considered psychopathological (Heber 1989). For example, Roman and Packer (1984) encouraged a right-hemispheric approach, based on popular notions of their time, as channeled by their 'spirit' guides Orin and DaBen:[5] 'Imagine all the cells in your right-brain, your receiving mind, reflecting perfectly the higher planes of reality, much like mirrors. Imagine the higher energy flowing from your right-brain into your left-brain, your conscious mind, with perfect precision and clarity' (Roman and Packer 1984: 76). And, with a nod to cryptomnesia: 'your guide... may take an idea you read about ten years ago, or use something you just learned yesterday... anything that is in your mind is a

---

5   Channeled entities are often solidified by using Biblical, classical or foreign-sounding names: Ramtha or Ramala, 'Ram' being Hindi for 'Lord of the Universe'. The name reflects the education of the creator. Jung, steeped in ancient associations, chose Philemon for his No. 2, Elijah for his wise old man, Salome for his anima.

potential tool for your guide' (Roman and Packer 1984: 51). They also recommended using a different tone of voice or an accent when chan-neling, suggesting a right-hemispheric provenance.[6] More scientifically, an fMRI study of a famous 'mentalist' performing a successful telepathic task showed 'significant activation of the right parahippocampal gyrus, whereas the unsuccessful control subject activated the left inferior frontal gyrus' (Venkatasubramanian 2008).

Roman first connected to her spirit guides through the Ouija board, a physical equivalent of the table-tapping alphabet system. After a traumatic car accident, she channeled 'Dan', then the 'higher vibrational' entity 'Orin' directly. She claimed to have suppressed Dan's true voice using her own instead.[7] Packer's entrée into dissociation occurred when he sensed a more knowledgeable presence assisting him while performing therapeutic body-work on clients. A PhD geophysicist by training, Packer resisted the New Age teachings until 'Orin' put him in touch with the 'higher perspective' of 'DaBen', complemented by his own philosophical, religious and scientific readings. Roman and Packer also reportedly developed a telepathic bond with each other. A close coupling or group concentration seems to be the *sine qua non* of mind-to-mind information facilitating 'spirit' contact and unconscious communication.

Fleck *et al.* (2008) studied 'transliminality', defined as the degree of transfer of unconscious thought and external stimulation into conscious awareness. Individuals with high transliminality would be predisposed to magical thinking, creativity, belief in the paranormal, as well as manic and/or mystical experiences. Previous studies of this type of individual had shown a leftward bias in attention (meaning looking to the left in attention from the right hemisphere), indicating hyperactivity in the right hemisphere; 'enhanced semantic priming for indirectly related words pre-sented to the left visual field, right hemisphere'; and a shift to right hemi-sphere complexity patterns. These individuals also displayed increased connectivity between the left and right temporal lobes, a phenomenon reported in NDEers, OBEers, and synesthetes as well.

The researchers used 'at rest baseline EEG' to measure enhanced right-hemispheric predisposition scientifically. They observed 'a reduction in beta power over the left posterior association cortex... coupled with a decrease in high alpha and low beta power in high-transliminality partici-

---

6   Léopold spoke with an Italian accent, although he did not speak or under-stand Italian.
7   Women channel male 'entities' to express 'otherness', a more authoritative voice or conflictual gender identity (Sword 2002; Platt 2007). Poetic muses spoke in the ear rather than possessed the body of the male poet. Modern male trance mediums, less common than women, channel male entities or speak as themselves in trance.

pants over a region of the right temporal lobe' (Fleck *et al.* 2008: 1359). The patterns of activation observed in high transliminality participants coincide with baseline EEG patterns observed in participants who show increased paranormal effects, *as well as* above average creativity levels. The researchers have theorized that differences within these groups may stem from a reduction in left-hemispheric lateralization for language and less internal synchronization within a hyperactive right hemisphere. Combined, baseline EEG changes may reflect differences in coherence or hemispheric integration, which may account for dissociative personalities, schizophrenic traits, unusual perceptions and creativity.

Unconscious Communication in the Therapeutic Dyad

Schore (2009) has said that psychoanalysis is 'undergoing a substantial reformulation from an intrapsychic unconscious to a relational unconscious whereby the unconscious mind of one communicates with the unconscious mind of another' (Schore 2009: 190). In the therapeutic relationship, psychoanalysis can restructure what was broken during the earliest relational experience. Both Schore and psychoanalyst Philip Bromberg proselytise for an 'empathic matrix' resembling a synchronous affective mother–infant bond, which can be easily repaired after subsequent disruptions (Bromberg 1998: 89). Bromberg, in particular, calls for therapist and patient to co-construct a safe, shared reality that recognizes internal 'otherness', as the therapist's self state shifts in tune with the patient's. As he enters into 'an authentic relationship with each voice' (Bromberg 1998: 200), the therapist helps negotiate new meaning in the patient's disparate self-narratives. Being accepted and dispelling shame allow the patient to 'express in language what he has had no voice to say' (Bromberg 1998: 16).

Jungian analyst Michael Conforti (1999) said that closely interrelating people, especially in a therapeutic dyad, can experience an increased neuronal charge and synchronization with each other. Two people can enter a single archetypal field constellating around 'self-similar/complex-similar interactions' (Conforti 1999: 81). If a patient has been sexually abused, the childhood trauma triggers the emergence of an archetypal field of abusive situations extending forward in time — unless the patient is released from its grip. A patient traumatized by maternal loss can exist in a field of abandonment until the pattern is recognized in a therapeutic break-through.[8] The weaker the individual's ego, the stronger the unconscious energetic field becomes, to the point of functioning autonomously.

---

8   As described, these energy fields sound like Rupert Sheldrake's morphogenetic fields. Sheldrake also firmly believes in telepathy, based on an animal-to-human model. He convincingly portrayed this at the 2008 Towards a Science of Consciousness conference with videos of a talking

In the process of healing, the therapist and patient can make inter-psychic connections. In a conference at the Jung Center in Houston, Conforti explained how one of his patients, who had been orphaned at birth and still felt traumatized by his loss, sought the mother in his analyst's psyche, recovering, in the process, personal information about her he had no way of knowing. Conforti asserted that unconscious information is constantly being shared in this manner. Atmanspacher *et al.* (2002) have similarly reported that a telepathic exchange of information between men-tally 'entangled' patients and therapists occurs in the phenomena of trans-ference and countertransference.

An Abused Therapist's Story

Annie G. Rogers, who had been severely abused as a child, later became a therapist for sexually abused girls. As an adolescent, she was labeled schizophrenic, manic-depressive or suffering from schizoaffective disorder at various times, but suspects she would now be diagnosed with DID. The presence of a guardian angel, 'Telesphorus', and several alter personalities who helped her deal with her trauma does suggest DID; but, the symptoms may overlap along a continuum of mental illnesses, as suggested by Claridge (1990) and Trimble (2007).

Despite yearly hospitalizations, medication and electroshock treatment blotting memories of much of her formal education, Rogers went on to become a child therapist and professor as well as a writer, painter and published poet. An unconventional therapist named Blumenfeld helped her access memories of childhood abuse, perpetrated by both parents, through dialog and dreamwork. But her own therapeutic relationship with a highly disturbed little boy she named Ben, who was abandoned by his mother and severely neglected by his foster family, made her repressed trauma conscious as the two became entwined in a single sphere domi-nated by loss, abandonment and abuse.

Rogers also read St. Augustine, Heidegger, Rilke and Virginia Woolf to understand her plight through self-similar association. Finally, Rogers discovered Freudian Jacques Lacan's theory of the language of the uncon-scious breaking through in repetitive syllables, consonants, vowels or homonyms to speak the 'unsayable' words of her patients' abuse history. Spotting these unconscious linkages became the means of bringing her patients' traumas to consciousness, often along with an artistic pursuit that allowed them to 'speak'.

---

parrot identifying pictures held by his owner in a different room and of a dog walking towards the door at the precise moment his owner walked towards her car to come home.

Rogers' poignant, poetic prose, like Judge Schreber's *Memoir of My Nervous Illness*, is invaluable for learning about dissociative processes. Whole phrases came to Rogers unbidden, as if from another person. In one instance, two lines from Shakespeare's *Measure for Measure* 'are spoken to [her], but not aloud: "Then, if you speak, you must not show your face. / Or, if you show your face, you must not speak"' (Rogers 1995: 68), which she interpreted within the context of 'unsayable' trauma. A poem she wrote to Ben came to her 'whole and formed, as if it were written by someone else' (Rogers 1995: 75); yet she is unable to account for hours of her day, an amnesia typical of DID. Disembodied voices prefaced a psychotic break where she was too wounded herself to heal another. During her hospitalization, angels visited, she dropped in and out of her body and 'words come into her mind as if from someone lost. She knows they are not really *her* words' (Rogers 1995: 103). An unidentified man morphs before her in distorted time and space. A voice in her left ear explains 'That's because we've disguised him' and show her a TV screen with more imagery; a voice in her right ear says 'Let it unfold' (Rogers 1995: 105). Later, as a therapist, Rogers affirms: 'trauma follows a different logic, a condensed psychological logic that is associative, layered, nonlinear, and highly metaphoric' (Rogers 2007: 54), or, we might say, right-hemispheric.

Rogers deteriorated into a paranoid schizophrenic state with boundary loss, thought disorder and somatic attempts at communciation. A 'body of light' comes through the walls and her own body becomes a 'body of light, tapping out messages in freezing and burning codes to unseen presences' (Rogers 1995: 108). A dissociative self moves alongside her, speaking to her as if to another person. Compelled to obey a 'chorus of unrelenting' voices, she remembers threatening to kill her former therapist-in-training, Melanie. As Melanie had become more professional and distant, refusing to 'mother' and touch her affectionately, Rogers fell prey to desperate measures, then slipped into silent madness.

As she resurfaced, her ability to speak and hear language slowly returned. She interpreted the broken shards of language usually labeled 'word salad' as valiant, yet incomprehensible, attempts at speech. The loss of language and tapped out coded messages suggest left-hemispheric breakdown and right-hemispheric somatic attempts at communication, as we saw in Elizabeth Mikal's case.

Blumenfeld brings her back through compassionate dialog and understanding, to the point where they 'understand one another's words and actions through [an] unconscious and powerfully deep connection' (Rogers 1995: 165). Their close therapeutic relationship will replicate hers with Ben. She wonders how 'this unconscious knowing passes from one human being to another' and 'if it depends on messengers', i.e. angels (256). Rogers' abuse history and mental breakdowns hone her skill at interpreting her patients' linguistic slips. As Blumenfeld says, 'You have a kind of giftedness, Annie, that probably has always been inseparable from your suffering... healing is always two-sided, isn't it?' (142–3).[9]

'Past Life' Regression and Unconscious Information

Dr. Brian Weiss

The unconscious grasping for self-sameness and two-sided therapeutic entanglement is also evident in the story of Dr. Brian Weiss, chairman emeritus of the Department of Psychiatry at the Mount Sinai Medical Center in Miami, and his life-changing experience with a young panic-stricken patient named Catherine. As detailed in *Many Lives, Many Masters* (1988), this 27 year-old woman started therapy for a debilitating anxiety disorder that was becoming progressively worse. Searching for possible childhood trauma behind her symptoms, Weiss found a depressive mother and an abusive, alcoholic father; neither this scenario, nor a current affair with a difficult married man, seemed sufficient cause for her distress.[10]

Hypnotized, Catherine described a terrifying scene where her father sexually abused her at age three. When Dr. Weiss suggested Catherine go back further, she described a series of 'past lives', all different, but with equally horrifying endings: drowning, tuberculosis, throat slashing or being sealed in a cave to die of leprosy. The moment of death itself always entailed floating above the body, going to an energizing light, meeting helping entities and passing on.[11]

Using Conforti's theory, Catherine seemed to attract stories of suffoca-tion, albeit retrogressively, or, using Gazzaniga's, she was wildly confabu-

---

[9] This unconventional treatment of a psychotic patient, paralleling R.D. Laing's approach, may have been successful because of the patient's training, reading influences and superior intelligence. It also fits with Carson's thesis regarding high intelligence as a protective factor in schizophrenia.

[10] Past life undercurrents ran through Rogers' story as well. In her childhood, she identified intensely with Joan of Arc, having received a 'command' from the Archangel Michael 'to find a way to translate the voices of angels for the world, to find words and an alphabet that would put an end to human suffering' (Rogers 2007: 6). During her recovery, she dreams that Nazis take her to be burned and hanged and sexually molest her as well. In another instance, she dreams that Ben is '[her] own child, come back to [her]', as if in a former life (Rogers 1995: 215).

[11] We see many NDE features in Rogers' mental relapse as well: the presence of light, floating above her body, angels and alters helping her cope. Near-ing death as well as struggling with unfathomable psychological devasta-tion bring on the same dissociative escapes — the mind's first line of defense against the fear of annihilation.

lating. The end scenario remained the same—the NDE model of a saving separation from the body. In either case, the spinning of successive 'past life' tales released her present anxiety's relentless grip on her throat.

During the intervals between her 'deaths', however, Catherine's demeanor and voice would change as the tone of the message became spiritual. She no longer described 'past' scenarios, but spoke, as though possessed, from the vantage of separate 'entities' called 'Master Spirits'. Each time the 'Masters' spoke, Catherine 'began to roll her head from side to side, and her voice, hoarse and firm, signaled the change' (Weiss 1988: 68). The messages were typical of the New Age: God is in each of us; there are different dimensions with higher levels of consciousness; we progress to higher levels through life lessons learned; we need to help others less evolved in successive lifetimes; we are sent back to new lives with increasingly greater psychic powers, talents and abilities' (Weiss 1988: 71). Debts as well as abilities are carried over.

Weiss was convinced that this information could not have come from Catherine. But the thoughts could have come from *him*, drawing on materials from his own knowledge and readings. In one instance, he asks her to 'see her life from a higher perspective... to answer her own question' about a memory of her real father's hitting her with a stick. Weiss had read about 'one's Higher Self or Greater Self' and used the idea in Catherine's therapy (Weiss 1988: 73). She responded that her father felt his children were intrusions in his life, that her brother had been conceived before the marriage (something she did not know, but was later confirmed by her mother). Dr. Weiss evoked Jung's collective unconscious to explain Catherine's 'superconscious' mind, her 'genius within', but later rejected this explanation (Weiss 1988: 74).

The 'Masters' tell the doctor that the messages are now meant for him, saying his own dead father and son are with them. The 'Masters' describe correctly how both died, suggesting again that the information was coming from Weiss. An unlikely adult cognition is proffered: the child had 'sacrificed' his life to absolve his parents' (unspecified) debts and to teach his father the 'limited scope' of medicine, in favor of psychiatry.

Convinced now of the existence of past lives, Weiss becomes a 'past life' therapist, claiming to have healed 3,500 patients by revealing their accumulated incidences of traumatic events or prolonged stressful life conditions. The success of his method might more likely be attributed to his patients *imagining* terrifying scenarios they transcended, along with assurance of death's impermanence. Ross's 'as if' scenarios also showed that *creating* alternative scenarios for actual past traumas can reframe or erase painful memories, as Epstein had in his Waking Dream Therapy. For Catherine, 'remembering' eighty-six past lives, with horrific death scenarios transcended, may have similarly deconditioned her phobias.

Catherine's ability to come up with specific, undisclosed information about her therapist may support Conforti's notion of a shared energetic field in a tightly connected dyad with the possibility of mind-to-mind contact. It could also suggest a bicamerality where the time-specific 'past' lives are confabulated in the left hemisphere and the timeless, disembodied Masters' voices, one of whom is a 'poet Master', are coming from the spiritually metaphoric right. Or, more mundanely, Catherine could have learned details about her psychiatrist's life from other doctors in his hospital where she worked as a lab technician.

Weiss remained convinced of the reality of Catherine's past lives and believed the whole experience had been *designed* to bring the message of reincarnation to a much wider audience. He added examples of scientists he knew who experienced the paranormal in different ways, but were reluctant to reveal to others: the warning voice of a dead father; solutions to research problems in dreams; visits to foreign cities that felt uncannily familiar. Yet, all of these are common instances of information brought to conscious awareness during a relaxed state as the 'Eureka' moment of revelation. Similarly, it was during sleep that he and Catherine became increasingly entangled as he awoke, at the same time as she, aware of her distress.

In an interesting corollary, Dr. Ross described the case of a man who came to him because of recurrent dreams about a former life in Greece and a strong sense of *déjà vu* on a trip to nearby Turkey. The patient, Bill, easily entered a trance state in Ross's office and presented as a boy in Didyma, Turkey. Answering Ross's questions about his location and names of his family members, the boy, at first vague, then provided Greek-sounding names, an alphabet and simple sentences. Ross said the language was inconsistent across time and clearly made up.

Within the trance locale, Bill brought forth a priest to talk to Ross. The conversation (in English) had the 'tone and feel of a rational interchange with an independent adult human being' (Ross 1994: 265). The priest, Charissos ('Ross' is cleverly scrambled to have a Greek ending), talks about the phenomenon of multiple personality in learned terms. Ross states, 'Bill's "unconscious mind" must have inserted the information about MPD because he knew that I would be interested in it. This was clearly theatre designed for me' (Ross 1994: 268). Possibly, Bill had dipped into Ross's mental reservoir to construct some of the scientific dialog. Ross recognized the mind's talent for creating fabulous fantasies, whether in the dream world, hypnotic trances or in a creative waking state.

Dr. Roger Woolger

In the above cases, one could ask if a mind-to-mind transfer is actually occurring or if these therapists and their patients were experiencing a *folie à deux*, or 'shared delusion'. Jungian past life psychotherapist Roger Woolger asked himself this question when introducing his skeptical entrée into past life regression in *Other Lives, Other Selves* (1988). In 1971, *The Journal of the Society for Psychical Research* had asked Woolger to review Arthur Guirdham's book, *The Cathars and Reincarnation*. Woolger learned of this French psychiatrist's mental entanglement with a female patient who had vivid, historically precise dreams about mass religious persecution in southern France during the thirteenth century.

At the time, Woolger diagnosed transference and countertransference in the patient and her therapist, especially when the two concluded they had been lovers during this sad period of French history. Eight years later, however, doing personal experimentation with a past life regression technique, Woolger vividly saw himself involved in the Cathar massacre, but as a brutal mercenary soldier rather than a victim. Further regressions subsequently showed him burned at the stake as a heretic — a horror he felt explained his lifelong fear of fire. He now understood his 'unbidden' violent fantasies and cynical rejection of Christianity and militarism. As in Helene Preiswerk's past lives, Woolger's patients' memories were full of sex and violence, abandonment and loss.

Woolger came to believe that the unconscious mind retains forgotten childhood memories from this life as well as from past lives. Intrauterine life, which depends on the attitude and emotional state of the mother, triggers 'karmic residues' as well (Woolger 1988: 264). Childhood abuse in this life triggers much worse memories from past lives. Woolger came to recognize past lives as 'other selves' lurking in the background of the psyche. The voices and visions of schizophrenics, in his scheme, were also past life fragments. Woolger, like Conforti, saw an archetypal pattern organized around painful past life fragments, resembling dissociative personalities with their different voices and mannerisms. Woolger added, '*that which is real for the patient*' is a psychic truth, not requiring irrefutable proof of reincarnation (Woolger 1988: 39), wondering if Jung's personality No. 2, Paracelsus and Goethe might have been past life fragments as well.

While far-fetched, Woolger's story does support the idea of unconscious communication in a therapeutic dyad. He began experimenting with a colleague interested in past life regression. Their dyad expanded to a group of six colleagues and friends working in pairs and meeting bi-weekly to share findings and conduct past life sessions. Three members of the group were 'omnivorous readers' who 'set out to read everything [they] could on past lives or reincarnation' (Woolger 1988: 19). Again, collaborating partners with broad readings could have brought unconscious information to the tightly knit group. Woolger's experiences bring to mind the Canadian Psychical Society's test case in which they created a biography of someone named Philip, whose 'spirit' a group of members 'conjured', despite the fact that he never existed at all (Owen 1976). Their belief and group concentration made the 'contact' work, just as requesting past life details in a therapeutic dyad can bring them out.

Stanislav Grof

Another convinced therapist was Czech psychiatrist Stanislav Grof, who first specialized in birth trauma regression using LSD-assisted psychotherapy, but now uses 'holographic breathwork' to induce altered states of consciousness. In his original drug therapy, he found that people 'reported "past life" experiences whose historical accuracy could later be confirmed. During their deepest sessions they were experiencing people, places, and things that they had never before touched with their physical senses' (Grof 1993: 17). His patients also showed a 'deep link with all those who have been abused, imprisoned, tortured, or victimized in some other way' (Grof 1993: 29). Correlating personal 'memories' of birth trauma to all forms of victimization may parallel the unconscious reaching for similar past life scenarios Dr. Weiss had found in his patients.

The Conscious Universe and Characteristics of Telepaths

Although not a therapist, consciousness researcher Dean Radin, a Senior Scientist at the Institute of Noetic Sciences with a Master's degree in electrical engineering and a PhD in psychology, has stated:

> …[W]e can get glimpses of information about other people's minds, distant objects, or the future or past. We get this not through the ordinary senses and not because signals from those other minds and objects travel to our brain. But because at some level our mind/brain is *already coexistent* with other people's minds, distant objects and everything else. To navigate through this space, we use attention and intention. From this perspective, psychic experiences are reframed not as mysterious 'powers of mind' but as momentary glimpses of the entangled fabric of reality. (Radin 2006: 264)

Radin claims everything in the universe is interconnected according to the laws of quantum physics. Since our brains are busily processing sense impressions in the waking state, we do not normally experience a deeper reality at that time. However, Radin adds, if the temporal lobes are unstable or shamanic methods like 'meditation, drumming, chanting, and psychoactive drugs' are used, something we have witnessed in the preceding chapters, entanglement resulting in paranormal experiences is possible (Radin 2006: 270). Further, certain types of people, like mystics and natural 'telepaths', can shift easily between normal and altered states of consciousness. Interestingly, the 'telepaths' he profiles are typically left-handed or ambidextrous, female, introverted, anxious, creative meditators, in line with Persinger's temporal lobe personalities and his enhanced right-hemispheric model (see Radin 2006: 44–9 for his 'characteristics of believers' and discussion of Persinger's experiments).

Based on a sensitivity questionnaire they developed, Jawer and Micozzi also found that being female, ambidextrous (implying using both hemispheres equally), appraising oneself as an introverted, imaginative thinker *and* recalling a plainly traumatic or series of traumatic events in childhood were major factors in those experiencing anomalous perceptions. Thin ego boundaries, the somatization of stress into illness, overexcitability, absorption and fantasy proneness play a role along with electrical sensitivity, synesthesia and migraines. Sexual frustration has long been associated with poltergeist effects in adolescents and mystical experiences in religious figures.

Jawer and Micozzi's model gets closer than Radin's to the truth of the matter when they assert that in order for anomalous events to occur, intense emotions and identity issues *must* be at stake. No embodiment, no consciousness. No crisis, no paranormal event. Citing the first law of thermodynamics, they connect their theory of frozen energy to reincarnation as well; while the entire personality does not transmigrate, the unreleased emotional energy does.

Continuing in this vein, they say: 'highly emotional events in a person's childhood can sensitize him or her to experiences that bear a certain resemblance — either concrete or symbolic — to the original occurrences. A remote event in space-time, according to de Graaf, if sufficiently momentous, may stand in for an experience in a person's own life, giving rise to such anomalous phenomena as crisis telepathy, precognition, or even poltergeist-like disturbances' (cited in Jawer and Micozzi 2009: 402). Similarly, Weiss's, Woolger's and Grof's patients might have also accessed a resonant life from the past, just not their own.

The similar emotions, perceptions and bodily states found in altered states of consciousness might be merely metaphoric, cryptomnesiac or they may point to nonordinary, non-local means of mind-to-mind contact. In any case, they remain effective insights for promoting new understanding and possible healing since the brain cannot easily distinguish 'as if' imagination from reality — as in dreaming, hypnosis and hallucinations. At the very least, the imaginative function of mind can use its story-making potential to construct scenarios that resonate with or replace the traumatic ones in the past.

Grof elected spiritual poets 'Rumi, Omar Khayyam, Kabir, Kahlil Gibran, Sri Aurobondo or Hildegard von Bingen' as best voicing 'the ultimate creative force' of cosmic consciousness (Grof 1993: 165). Poetry and the paranormal remain entwined. Radin cited English poet Francis Thompson: 'All things by immortal power, / Near and Far / Hiddenly / To each other linked are, / That thou canst not stir a flower / Without troubling of a star' and ends his book with the well known verse of William Blake: 'To see the world in a grain of sand / And heaven in a wild flower, / Hold infinity in the palm of your hand / And eternity in an hour'.

***

A sensed presence or speaking 'other' is a real, that is, subjectively experienced phenomenon. Welcomed in primitive times as a sign of special election and a source of knowledge, in modern times, dissociative breaks are deemed pathological and treated as such. The nineteenth century maintained a keen interest in the newly discovered 'unconscious', tracing hysterical, dissociative symptoms to heritable conditions and/or early childhood trauma. Highly intelligent theorists, such as Jung, Kerner and Myers, with right-hemispheric proclivities themselves, remained open to the possibility of information transmitted via unconscious means and even the survival of death.

Recent scientific research shows how genetic traits combined with triggering traumas can produce dissociation along with observable changes in the physical brain. Since the brain develops in relation and damage often occurs in the wake of negative relational experience, healing is best effected in relation as well. While the therapeutic process itself can create additional splitting, in an effective therapeutic dyad where patient and therapist share mental states, surprising revelations can occur.

Healing a damaged mind requires empathy and superior intelligence. Radin signaled Carson's earlier research showing that people with a low latent ability to inhibit confusing sensory input from the outside are more predisposed to mental disorder. Following a study of Harvard University students, he said that the most highly creative students had 'both lower latent inhibition and higher IQ as compared to the other students' (Carson *et al.* 2003). Their lowered latent inhibition allowed for creative insights and associations that others might not perceive, while high intelligence protected them from falling into the abyss of madness that could consume lesser minds.

Therapists such as Ross, Conforti, Weiss and Woolger showed how an empathic engagement with the imaginative constructs of their patients, treating them 'as if' real, allowed for nonlinear, self-similar, resonant attraction that could rewrite or rewire maladaptation into healthy assimilation. Whereas the therapists' beliefs, expectations and even unarticulated knowledge could impact their patients, the reverse was true: the therapists could be converted to their patients' beliefs, as were Weiss, Woolger and Grof.

A therapist can bring out troubled patients' stories, whether fact or fantasy, to explain the roots of their pain. Poets, artists, musicians and mystics often forge their creative constructs on the anvil of their own suffering. In so doing, they earn lasting recognition as geniuses of invention whose heightened perception of reality can benefit others for whom it rings true.

Chapter Five

From Myth to Mediumship: John Keats and Victor Hugo

He ne'er is crown'd / With immortality, who fears to follow / Where airy voices lead…
  — Keats, Endymion: Book II, ll. 211–13

Ainsi Nature! Abri de toute creature / O mère universelle! Indulgente Nature. [Thus Nature! Shelter for all creatures / O universal mother! Indulgent Nature.]
  — Hugo, Les Chants du crépuscule

In the face of difficult problems or in answer to serious questions, the voices of immaterial beings can constellate from the personal or cultural repertoire of the minds that create them. They may retrieve hidden memories, make novel connections, or make illuminating associations from their immediate surroundings. At the outer edges, they may prophesy the future or borrow information from another's mind. Dissociative constructs tend toward the concrete and the visual — symbols, archetypes, parables, metaphors, neologisms, geometric schemes and designs, still or movie-like imagery — as they seek a larger context to frame their answers. Both in form and in content, the messages show the right hemisphere's involvement. Poets, who already veer toward right-hemispheric linguistic expression, are more likely to experience dissociative creativity. Just as personal trauma produces dissociative identity disorder (DID), societal periods of crisis are more likely to trigger eruptions from the right who command, counsel, warn and heal.

Sensitive, right-enhanced, Romantic poets rejected left-hemispheric Enlightenment logic and the Industrial Revolution's machine-oriented world. The confluence of childhood trauma and societal-wide ills triggered a poetics of revolt, forged not only by personal loss and separation, but also by a collective yearning for a consolatory return to Mother Nature. Especially in France, conventional, classical form gave way to a bold new vision that rejected old rules, striking off on a liberating fugue towards self-expression.

At the most fundamental level, McGilchrist differentiated the left hemi-sphere's focus on our 'narrow... needs' versus the 'broad, open... world apart from ourselves' (McGilchrist 2009: 27). He also cited Schore's right-hemispheric mother–infant paradigm as evidence of the right hemisphere's function in creating a solid sense of self, differentiated from others. Here, he delineates the damage:

> Damage to the right parietal and medial regions [of the brain] may result in confusions of self with other; damage to the right frontal lobe creates a disturbance of ego boundaries, suggesting 'that the right hemisphere, particularly the right frontal region, under normal circumstances plays a crucial role in establishing the appropriate relationship between the self and the world'. It is this region that is so obviously not functioning properly in schizophrenia, where subjects not only lack empathy, humour, metaphorical understanding, pragmatics, social skills and theory of mind, but crucially mistake the boundaries of self and other, even at times feeling themselves to melt into other individuals or that other beings are invading or occupying their own body space. (McGilchrist 2009: 90)

Keats and Hugo were not schizophrenics. They neither melted into others nor lost their *corporeal* sense of self; but they did experience either *mental* merging or splitting. Keats understood the Other from the inside. Hugo proliferated external Others, all aspects of himself, who liberated his creative thoughts and authorized him to express what he already knew to be true. Their dissociative tendencies oiled the engines of their metaphoric fluency.

Romanticism

As the ultimate apologist for the right hemisphere, McGilchrist said Romanticism exemplified a right-hemispheric 'grappling with experience which is multiple in nature, in principle unknowable in its totality, changing, infinite, full of individual differences', as opposed to the left hemisphere's 'simple, knowable, consistent, certain, fixed therefore ulti-mately finite' view (McGilchrist 2009: 352-3). Reducing the dichotomy further, he said, 'Romanticism is a manifestation of right-hemispheric dominance in our way of looking at the world' (McGilchrist 2009: 354). It 'redresses the imbalance of the hemispheres... and curtail[s] the dominion of the left' (358).

The Romantics certainly reacted against the eighteenth-century rationalist philosophers who had emptied the mind of a God, filling it with science alone. The Romantics wanted to retain a place for 'natural super-naturalism', which, in effect, elevated the creative imagination to divine status (Abrams 1971: 66, in Armstrong 1993: 347). Historically, the poet as prophet harks back to the ancient period described by Jaynes, when reli-gion was expressed as poetry and poets were purveyors of divine truths. Yet, in the nineteenth century, Hugo still stated categorically, '*le poète est prêtre* [The poet is a priest]' (Hugo 1864/2003: 75) and that God creates through great minds. We now understand how both poets and mystics use the novel, intuitive, emotional, metaphoric resources of the right hemisphere in their creative processes. Metaphorically, I would add, Romanticism is a matriarchal mentality—a return to the Mother—where truth and beauty are found in unbridled Nature, rather than in patriarchal societal constructs.

The Romantics themselves probably did not realize that a genetic predisposition to mood disorders, along with external circumstances, childhood trauma being paramount, produced their 'God-like', i.e. dissociative linguistic genius. Other factors promoting dissociation include a lonely, fantasy-prone personality since childhood, with imaginary companions (Murphy 1993, Ross 1989); latent homosexuality, fear of loss of control, sensitivity to humiliation (Baker 1996: 292); parental divorce, untimely death of a mother, and intentional exile (Brown 1997). Under these conditions, the right hemisphere might take control to restore emotional homeostasis, with messages that *feel* or *sound* like a divine or authoritative other. At times during the séances, Hugo and Shakespeare co-write a poem, commenting on each other's contributions.

John Keats

In 1816, Keats passed his examinations to become a doctor; but, feeling he possessed 'Abilities greater than most Men', he chose what he considered a higher calling: a poet. Isolating himself from his friends, he read other poets, especially Milton, Wordsworth, Dante and Shakespeare, and classical legends. Imagination, for Keats, was sacred: 'I am certain of nothing but of the holiness of the heart's affections and the truth of the imagination—what the imagination seizes as beauty must be truth—whether it existed before or not' (Letter to Benjamin Baily, 22 November 1817, in Keats 1990: 365).[1] 'Negative Capability' expressed the idea that 'uncertainties, mysteries, doubts' could be held in abeyance, 'without any irritable reaching after fact and reason' (Letter to George and Thomas Keats, 21 December 1817: 370). As Armstrong put it so well: 'Reason had only a limited role in this creative process… Like a mystic, the poet had to transcend reason and hold himself in an attitude of silent waiting' (Armstrong 1993: 347). The poet waited, I would say, for the associative shards of non-conscious materials to coalesce into meaningful linguistic expression. Wordsworth made poetry from his own epiphanic experience. He had 'see[n] into the life of things' and experienced 'that serene and blessed mood' where time stands still in an all-encompassing sense of

---

[1]   All subsequent Keats quotes reference this edition, unless otherwise indicated.

'harmony' and 'joy' ('Lines Composed a Few Miles Above Tintern Abbey', *The Prelude* II: 256–64).

Keats's personality and life history fit the profile of those susceptible to right-hemispheric illuminations. He was described as a 'violent and ungovernable' child and classified as bipolar in *Touched With Fire* (Jamison 1993: 70). He described his own depressive state as 'sitting with your wings furl'd for six Months together' (Letter to Percy Bysshe Shelley, 16 August 1820, cited in Jamison 1993: 323). Wings are a metaphor for both higher creativity and angelic visitations. In his poem *Lamia*, Keats bemoaned cold (left-hemispheric) philosophy silencing (right-hemispheric) mystery: 'Philosophy will clip an Angel's wings, / Conquer all mysteries by rule and line, Empty the haunted air, and gnomed mine– /Unweave a rainbow'. He found company in Nature: 'The roaring of the wind is my wife and the Stars through the window are my Children' (Sperry 1973/1994: 46). But his losses were insurmountable–his father at eight; his mother's remarriage two months later; her death when he was fourteen. He had also lost a baby brother and his beloved grandmother. Compounding those losses were tuberculosis, an inability to marry his beloved because of poverty and, finally, his self-imposed exile in Italy for worsening health.

Keats constructed a 'system of salvation', molded, perhaps, as a rationalization for his enormous suffering. With *imagination* considered sacred, not surprisingly, the Christian concept of the mediatory role of a crucified Jesus did not work for him. Salvation came through the trials and tribulations of one's own suffering. Among the letters written to his brother and sister-in-law, George and Georgina, who had gone to America and were sorely missed, Keats had elaborated: this world is not 'a vale of tears', but 'the vale of Soul-making'. Intelligence is a divine spark, an 'atom of perception' that must pass through this world of woe to gain personal identity as a soul. The Heart is the centerpiece of a triptych framed by Intelligence and World: 'Do you not see how necessary a World of Pains and troubles is to school an Intelligence and make it a soul? A Place where the heart must feel and suffer in a thousand diverse ways!' In his system, the Heart is both 'Hornbook' and 'Mind's Bible' and the 'teat from which the Mind or intelligence sucks its identity' (Letters to G. and G. Keats, February, March, April, May 1819: 473–4). In this curious formulation, Keats drew metaphorically on what he ostensibly *lacked*–poetic knowledge, the Mother and an identity–all of which would be remedied by reading the great poets, merger with Mother Goddesses and becoming a great poet himself.

Keats, unlike the other poets we will study, had a tendency to merge, rather than split, calling himself the 'chameleon Poet' (Letter to Richard Woodhouse, 27 October 1818: 419). Keats exemplified Apollonian deification. Annihilating the self, he fused with the other. Yet, this capacity for self-annihilation permitted a deeper knowing of others, an ability to enter their skin and describe them from the inside. By Keats's own admission, he lost his identity in a crowded room and left with everyone else's inhabiting his psyche. This merger indicates right-hemispheric hypersensitivity at work, resembling a dissociative state in which the self disappears to become another person, or, as with mediums in a trance, the self assumes another's past life. In 'Endymion', Keats depicted merger with a Goddess as marriage to the Muse, from which effortless creativity flows, entraining immortality.

Keats's creative process did, in fact, favor associations and images over labored analysis (see Sperry 1973/1994). After witnessing him at work, a member of his circle wrote: he 'sat down to his task, – which was about 50 lines a day, – with paper before him, & wrote with as much regularity, & apparently with as much ease, as he wrote his letters' (in Rollins 1965: 270). It is also easy to see how another person's poetry could wind its way into his dreamlike creativity. In a 'rather low state of mind', after reading the fifth Canto of Dante's *Inferno*, Keats dreamt he was in the second circle of Hell. Whereas Dante's adulterous lovers had been tormented by great gusts of wind, Keats's imagination landed on the tale of Francesca and Paolo, engaged in a furious kiss after reading the story of Lancelot and Guinevere's forbidden love. An associative concatenation of stolen kisses from the Middle Ages led to a beautiful dream, refashioned as a sonnet which will betray the dream's actual feeling:

> The dream was one of the most delightful enjoyments I ever had in my life – I floated about the whirling atmosphere as it is described with a beautiful figure to whose lips mine were joined at it seem'd for an age – and in the midst of all this cold and darkness I was warm – even flowery tree tops sprung up and we rested on them sometimes with the lightness of a cloud till the wind blew us away again – I tried a Sonnet upon it – there are fourteen lines but nothing of what I felt in it – o that I could dream it every night. (Letters to G. and G. Keats, in Keats 1990: 469).

The long-lasting kiss of the dream remained 'sweet' in the sonnet, but the lips became 'pale', less wish fulfillment than a sense of foreboding: 'Pale were the sweet lips I saw, / Pale were the lips I kiss'd, and fair the form / I floated with about that melancholy storm' ('A dream, after reading Dante's Episode of Paola and Francesa', ll: 10-14). The playful drifting of the dream lovers could exemplify effortless creative flow, while the pale lips embodied a more tortured relationship to the Muse. A few days later, that negativity culminated in 'La Belle Dame Sans Merci', in which a 'knight at arms / Alone and palely loitering' remained transfixed on the hillside where he fell sway to a 'faery's song'. Fairy-dust brought fairy-lust, then abandonment on a cold hillside. But first, the knight dreams of 'pale kings and Princes too / Pale warriors, death pale were they all; / They cried "la belle dame sans merci / Thee hath in thrall"' (ll. 37–40). In the act of writing, the poet converted pleasure into pallor, falling prey to negative right-hemispheric emotions. Perhaps the 'belle dame' is not only Robert Graves' White Goddess or Camille Paglia's oppressive Mother Goddess, but also a vestige of the personal mother, gone, yet relentlessly beckoning the poet to recreate or rejoin her.

Both poet and physician, Keats filled his poetry with references to the brain. As he said in 'The Fall of Hyperion' (1819), the 'poet is a sage; / A humanist, Physician to all Men' (Canto I, ll. 180–90: 295). The poet illuminates 'dark secret Chambers' in striking metaphors, we might say, outshining the fMRI images of today's neuroscientific scans (Canto I, l. 278: 297).

Reading the Two Hyperion Poems

Dreams, as Freud, Jung and Tibetan Buddhists agree, are potent bearers of messages from the unconscious. The Romantic poets also recognized the dream state as a source of inspiration. Recalling that a dream character commanded me to read the two 'Hyperion' poems, let us explore their connection to my friend's 'angelic' voices. In 'Hyperion: A Fragment' (1818), a Goddess comes to the dying Saturn in his sleep. 'In solemn tenour and deep organ tone' she speaks directly into his ear, reminding me of the booming, guttural voice of my dream character (Book I, l. 48: 226). Sorrowful, weeping at Hyperion's feet, she is nonetheless unconsoling. The winds and waves are silent before fallen divinity. As Jaynes recognized, dissociative voices came to the ancient Greeks as the environmentally attuned right hemisphere converted the sound of the wind and waves into speech. Keats expressed another recognition of the poet's power to convert Nature's voice into human speech in the 'Fragment. Where's the Poet? Show him! Show him!' in which the 'the Tiger's yell / Comes articulate, and presseth / On his ear like mother-tongue' (ll. 13–15: 224).

In 'Hyperion: A Fragment' (1818), the Mother Goddess tells Saturn he must accept his demise, giving way to the New. As the sluggish Saturn awakens, he recognizes the Goddess Thea, Hyperion's wife. Like an infant seeking confirmation in his Mother's face, he pleads she open her eyes: 'I am gone / Away from my own bosom; / I have left my strong identity, my real self' (Book I, ll. 112–14: 228). He asks the Goddess to open her eyes and 'sphere them round', searching all space for shapes like wings or chariots, the metaphoric messengers of myth and religion, 'to repossess a heaven he lost erewhile' (Book I, ll. 117–24: 228). Aroused, with hope renewed, Saturn turns back into the woods to rally others to the cause.

Meanwhile, Hyperion, still God-like but insecure, is the last remaining Titan. Fear, horror and darkness stew in the right hemisphere's cauldron of negative emotions: 'Dreams', 'spectres', 'phantoms' and 'shady visions' haunt him. Coelus, his father's voice from Heaven, taking pity, 'whisper'd low and solemn in his ear', reminding his divine son that he still has freedom of movement. The commanding Coelus, an 'ethereal presence', lacks embodiment: 'I am but a voice; / My life is but the life of winds and tides, / No more than winds and tides can I avail' (Book I, ll. 306–42).

The awakened Saturn invokes Oceanus, God of the Sea, for assistance. Murmuring his counsel, Oceanus speaks of eternal laws to comfort his Titan brother: 'We fall by course of Nature's law, not force / Of thunder, or of Jove' (Book II, ll. 181–2). As self-organizing principles of the nervous system, the gods speak broadly in terms of laws of the universe, the largest contextual correlate of the human mind. As Chaos and Darkness engendered Light, Nature's law requires 'fresh perfection', ever evolving towards newness and beauty (Book II, l. 212: 239). In the silence following his counsel, the voice of his daughter Clymene recounts how she once heard the name 'Apollo' hovering around like a dove 'with music wing'd instead of silent plumes' (Book II, l. 287) — a word heard on an environmental flutter. The booming voice of Enceladus overrules Clymene's, as he incites the fallen Titans to action, brandishing the image of Hyperion, God of the Sun: 'Speak! roar! shout! yell! ye sleepy Titans all' (Book II, l. 316: 242). Book III deserts the fallen Titans to recount the tale of Apollo, god of music, poetry, healing and prophecy, destined to replace Hyperion. Clustered by the Greeks, these domains can now be recognized as right hemispheric.

In keeping with the 'unwilled' dissociative theme, Keats told his friend Woodhouse that Apollo's speech 'seemed to come by chance or magic — as if it were something given to him' (in Keats 1990: 594 n. 246). Apollo weeps with 'half-shut suffused eyes' when '[w]ith solemn step an awful Goddess came' (Book III, ll. 44–6: 245). As he plays the lyre which this Great Goddess of memory, Mother of the Muses, had placed by his side while he slept, 'all the vast / Unwearied ear of the whole universe / Listen'd in pain and pleasure at the birth / Of such new tuneful wonder' (Book III, ll. 64–7, 83: 245–6). Apollo pronounces her name without knowing how: 'Mnemosyne!'

The enchantress, about whom he had dreamed, he now beholds in person, like Jung his Philemon. Having recognized and named his Goddess in a manner akin to telepathic knowing, Apollo asks her many questions. With speech unnecessary, the response explodes in a panoramic flash, like Swedenborg's angelic 'thought balls', Monroe's extra-terrestrial 'rote balls', or the compressed learning technique of the humans in the Matrix. All knowledge pours into his brain, thus 'divinizing' him:

> Mute thou remainest — Mute! yet I can read
> A wondrous lesson in thy silent face:
> Knowledge enormous makes a God of me.

Names, deeds, gray legends, dire events, rebellions,
Majesties, sovran voices, agonies,
Creations and destroyings, all at once
Pour into the wide hollows of my brain,
And deify me, as if some blithe wine
Or bright elixir peerless I had drunk,
And so became immortal. (Book III, ll. 112–20: 246)

The fragment halts with Apollo convulsing in shock, quivering and speech-
less, dying into divine life through awareness of suffering, as the Goddess
lifts her arms in the ancient pose of prophesy.

In 'The Fall of Hyperion: A Dream' (1819), the Keats-like poet merges
with Apollo, in a manner both privileged and precarious. First, he
comments on how ancient mystics and primitives 'Guess at Heaven'
without writing down their words, since they are not poets. (*While they may
have 'guessed at heaven', they were almost always poets, parable makers or
storytellers, who reached their audience though trance-like 'enchantment', while
others wrote down their words.*) Yet, with foreshadowing or precognition, he
opines, since 'every man' has visions, the judgment on whether he writes
as poet or as religious fanatic 'will be known / When this warm scribe [his]
hand is in the grave' (Canto I, ll. 1–18: 291).

After eating and drinking a mysterious meal left within some beautiful
woods, the poet swoons and falls asleep, awakening in an altered landscape (and, no doubt, state of consciousness). Architecture, vestiges and
vestments of ancient gods lie about him. In the distance stands… an altar,
towards which he 'sober-pac'd' approached (Canto I, l. 93: 293). He sees a
sacrificial fire, enshrouded in clouds of bliss-inducing forgetfulness. A dire
threat is voiced from within the sacred fumes:

If thou canst not ascend
These steps, die on that marble where thou art.
Thy flesh, near cousin to the common dust,
Will parch for lack of nutriment — thy bones
Will wither in few years, and vanish so
That not the quickest eye could find a grain
Of what thou now art on that pavement cold. (Canto I, ll. 107–13: 293)

A 'palsied chill' ascends his limbs, nearly suffocating him, until 'One
minute before death, [his] iced foot touch'd the lowest stair; and as it
touch'd, life seem'd / To pour in at the toes…' (Canto I, ll. 121–34: 294).
Surviving his near death experience, he asks the High Priestess to 'purge
off… [his] mind's film', thus opening the doors of perception, recalling
Blake's axiom.

From behind her veil, the sad Goddess Moneta says: 'the scenes still
swooning vivid through my globed brain / With an electral changing
misery / Thou shalt with those dull mortal eyes behold, / Free from all
pain, if wonder pain thee not' (Canto I, ll. 244–8: 297). The veil now parted,

her blanched, death-like countenance nearly frightens off the poet, but for the 'benignant light' from her half-closed, inward looking eyes. He aches to 'see what things the hollow brain / Behind enwombed: What high tragedy / In the dark secret Chambers of her skull / Was acting, that could give so dread a stress / To her cold lips, and fill with such a light / Her planetary eyes; and touch her voice / With such a sorrow' (Canto I, ll. 277–82: 297–8). His aching brings her by his side; she whispers in his ear; and the sad tale of Saturn unfolds again.

This time, the poet does not gain his knowledge all at once, but rather by entering her head and seeing *through her eyes*, 'as a God sees': a merger, not an instantaneous illumination. In Canto II, Moneta begins her tale of Hyperion, explaining she must 'humanize my sayings to thine ear; / Making comparisons of earthly things; Or thou might'st better listen to the wind, / Whose language is to thee a barren noise, / Though it blows legend-laden through the trees...' (Canto II, ll. 1–6: 302). The Gods and Goddesses must use METAPHOR to elucidate; hence poetry is born as the language of the Gods.

The Apollonian poet is not 'feminine because passive to his own vision', or 'haunted by daemonic hierarchic females', as pronounced by Paglia (1990: 329, 388). Nor is the poet's emergent 'feminine consciousness' an escape from patriarchy's stranglehold that may have 'precipitated his early death' through excessive sensibility, as Marion Woodman (1990: 15–16). I would say that monumental losses bred mythic mothers informing him through a breach in unified consciousness his own suffering had created. Keats's poetic espousal of mythic tales—their voices, visions, dream encounters, mergings and metamorphoses—shows not only his own right-hemispheric proclivities, but also the ever-present dissociative conscious-ness of his ancient poetic forbearers. Keats's mythopoeic exploration of suffering, death and renewal did not kill him, but may have arisen from an intuitive or precognitive sense that he would shortly die himself.

In the great odes, Keats relied more on the workings of the conscious brain, less on a dissociative creative impulse: 'Yes, I will be thy priest, and build a fane [temple] / In some untrodden region of my mind, / Where branched thoughts, new grown with pleasant pain, / Instead of pines shall murmur in the wind' ('The Ode to Psyche', ll. 50–3: 280). Yet, his verbal fluency would remain undiminished. Keats's friend, Brown, remarked in a letter that he saw the poet write 'Ode to a Nightingale' in just one morning. Sperry (1973/1994) considered the great odes "a series of closely related and progressive meditations on the nature of the creative process, the logical outgrowth of his involvement with Negative Capability' (243).

The message I gleaned from the 'Hyperion' poems is that voices are consolatory in times of crisis and that the face of the Mother Goddess, and particularly her eyes, transfer unconscious knowledge directly to her charge. Receiving the knowledge of immense suffering, transcending time and space, overwhelm the mortal; it is a sacrificial boon that confers 'divinity', that is, anomalous knowing. Keats's suffering spawned his poetry, making a Soul of his divine spark. He died too young, but his words remain immortal. In 'Endymion', he reminded us that 'A thing of beauty is a joy for ever' (Book I, l. 1: 61). 'Ode on a Grecian Urn' pronounced his final truth: 'Beauty is truth, truth beauty, — that is all / Ye know on earth, and all ye need to know' (ll. 49–50: 289).

Victor Hugo

While Keats merged Apollonian-style, Hugo veered toward Dionysian fragmentation. Hugo's poetic grandeur seemed to be in direct proportion to the traumas he suffered as a child. He was born on 26 February 1802, thirteen years after the beginning of the French revolution, when the government had already reverted to monarchy and an empire. Thus, he entered the world in a time when battling politico-social realities would be the norm for most of his life. Hugo's father, then a major in Napoléon's army, had wanted a girl when Victor was born, the third of three sons. A sickly newborn and predicted to die, Hugo was still unable to hold his head upright at fifteen months. Despite his ill health, Hugo's family moved to Marseille six weeks after his birth. His mother, Sophie, left her children at the end of November to return to Paris. Victor was only 8 months old when his mother left, not returning until he was 17 months old, the crucial period for establishing a secure sense of self (Schore 1994). Abandonment, humiliation, parental separation and exposure to the horrors of war marked his entire childhood, not without profound effect (biographical details in Robb 1997/1998).

Hugo's parents officially separated when he was 12 years old; his father then sent him and his brother, Eugene, to a *pension* in Paris, with only a stern paternal aunt to guard and spy on them. Victor considered his father 'the agent of "Implacable Destiny" who kept him apart from his sainted mother' (Robb 1997/1998: 54). Although Hugo adored his mother, she was no saint. She was very strict, once forcing him to dress as a girl for crying when he was only 5 years old. Because of her stubborn objection, he could not marry Adèle Foucher until after his mother had died.

Robb's detailing of Hugo's years as a young student in Paris give us some indications of his brain lateralization. He despised mathematics (left-hemispheric), while performing admirably in visuospatial geometry (right-hemispheric). He preferred using imagery and intuition to linear calculations when problem solving. During his years at the *pension*, Hugo wrote hundreds of lines of verse, a play, fables, song lyrics and memorized thirty lines of Latin at bedtime, translating them into rhyming couplets when he awoke (Robb 1997: 70). With poetry constantly on his mind, he awoke with 'complete alexandrines and even whole poems' forming in his head, committed to paper upon awakening (Robb 1997: 55). His incredible facility with practically unconscious verse composition will play out again during the exile séances on the isle of Jersey.

Were his poetic outpourings at the Parisian *pension* a conscious effort to avoid the military destiny his father had preordained, as Robb contended? Possibly, but I would add that his traumatic early childhood and present separation from his mother were now behind the impulse. The poetic prizes he garnered, including from the *Académie française*, gained his mother's esteem as well.

A genetic predisposition to emotional instability, so evident in other great poets, was apparent in Hugo's family as well. His brother, Eugène, who was writing poetry at the same time, was less fortunate, suffering a mental breakdown on the day that Hugo married Adèle. Where positive affirmations and collaborations protected Hugo, other family members would succumb to mental illness. Eugene would die fifteen years later in an asylum, bringing to mind van Gogh, who also suffered a breakdown the day his brother, Theo, got married. Loss of the beloved brother to marriage may have seemed tantamount to abandonment. Hugo's daughter, named Adèle like her mother, was confined to an asylum as well, after having searched hopelessly for her English lieutenant colonel who had traveled to the New World. Internment in an asylum had grim prospects in nineteenth-century France, as occurred with Paul Claudel's artist sister, Camille, and Maupassant, as we saw.

There are other external signs of Hugo's inner mind. First, he appeared to be left-handed, for the most part, even though he wrote with his right, as compelled to do in the French school system. In photos, we most often see his left hand supporting his head in a typical Hugolian pose. In other photos, with his arms across his chest, Hugo's left hand was up. He also parted his hair on the right like a left-hander. He was an artist as well as a poet, as seen in the emotionally evocative ink paintings from his exile period and the exquisite hand-made decorative items in his exile house on Guernsey. Art Critic James Hall (2008) said that the highly prolific Hugo, in art as well as in literature, 'cultivated ambidextrousness', drawing in ink with his *left* hand to access the unconscious. Hall described Hugo's 'ink drawing *The Dream* [as] consist[ing] of an energized yet disembodied left arm, shirt unbuttoned at the cuff, thrust upwards with fingers outstretched and palm open towards a turbulent, smoky sky, as if demanding to become a lightning-conductor for cosmic electricity' (Hall 2008: 338).

Beyond his art and poetry, Hugo was prone to paranormal beliefs, as we have come to expect in right-enhanced minds. Hugo claimed the wind and waves 'spoke' to him, not metaphorically but actually. Extremely sensitive, he did not allow cut flowers in the house believing they suffered when being cut. He believed two dogs that followed him around his island home were the reincarnated souls of repentant political enemies. He saw ghosts. He had frightening dreams where the dead spoke to him. Mysterious lights and knocking sounds troubled him in his Marine Terrace house on Jersey. Hugo was said to have had an extremely active magnetic field; when he was deprived of sexual activity, 'supernatural' sounds could be heard throughout the house (Robb 1997/1998).

Neither baptized nor adhering to a religion, Hugo was nonetheless 'profoundly religious' (Gaillard 1981: 10), suggesting he could be categorized as one of Ramachandran's and Persinger's right temporal lobe personalities. De Mutigny (1981: 35-6) said that Hugo believed in God intuitively, as a vague and cosmic being. Hugo abhorred specific religions and their dogmas, which he felt separated and persecuted people in equal doses. He could not fathom how a just God could punish an innocent, like his daughter, Léopoldine, who died at 19 in a boating accident, along with her new husband Charles. Hugo would find some consolation in the notion of metempsychosis, where sins in a previous life are punished in a subsequent one. Taken together, genetic predisposition, childhood trauma, the death of loved ones, especially his beloved daughter; then political exile after Louis-Napoléon Bonaparte's coup, led to a mental state amply primed for the collaborative séances—between friends and family as well as with the conjured minds of great poets past—that would occur on Jersey.

Hugo's close friend, Auguste Vacquerie, brother of the young husband who lost his life in the capsized boat with Léopoldine, had followed Hugo into exile and was thoroughly engaged in the séances. A journalist, playwright and published poet himself, Vacquerie wrote *Miettes de l'histoire*, a book about their experiences and the culture-clashing Franco-Anglo atmosphere on Jersey. He explained how Delphine de Girardin, a rich journalist and poet who maintained an important literary salon in Paris, visited Hugo's exile home for ten days in September 1853.

Séances using the 'talking tables' (*les tables parlantes* or *tournantes*) had become all the rage in French society. F.W.H. Myers and his cohorts in England were also exploring the existence of life after death, pursuant to the Fox Sisters' experience in the United States (see Hamilton 2009). Mme de Girardin, who was ill herself and would die of cancer the following year, had lost someone close the previous year (Vacquerie 1863: 55). With her interest in the phenomenon peaked, she suggested the assembled friends and family try to contact the dead. Mme de Girardin would then go back to Paris and spend the last year of her life talking with 'spirits'— including Mme de Sévigné, Sappho, Molière and Shakespeare—through the tables. She died among them, '*sans résistance et sans tristesse* [without resistance and without sadness]' (Vacquerie 1863: 409).

Putting a three-legged pedestal table on a larger table, two people placed their hands on top, waiting for a response to questions posed by a séance participant. Despite their efforts and the conducive atmosphere, nothing happened until the next to last evening of Mme de Girardin's stay. Suddenly, the table began to move, tapping out messages letter-by-letter ('a' = one tap, 'b' = two, etc.; 'yes' = one tap, 'no' = two). At first, the table correctly 'guessed' what word or incident participants were thinking of, to the participants' astonishment. The table then refused to answer any more 'absurd' questions. 'Who are you', Mme de Girardin asked. 'Dead girl', 'Ame soror', the 'spirit' replied. Général Le Flô, another exile, asked, 'Whose sister is it? Charles [Hugo's oldest son] and I have both lost a sister'. Everyone thought of Hugo's lost daughter and cried. In response to Hugo's pointed questions, the dead girl replied that 'yes' she was happy; she was in the 'light'; and expressed the need for others to 'love', if they wished to go to her (see Simon 1923/1996: 9–15).

Victor Hugo claimed his son, Charles, was the medium for the tables, because magnetism 'mulitplied one's intelligence five times'. The 'spirits' also deemed Charles the real medium and insisted on his presence. The well-read poet James Merrill would refer to this same formula in his own poem, *The Changing Light at Sandover*, saying his partner David was the 'hand' for their Ouija board odyssey. While Hugo rarely put *his* hands on the table, he asked questions and often wrote up the transcripts of the proceedings or reviewed those transcribed by his daughter, Adèle. For the next two years, the table would rarely respond unless both Charles and his mother were holding the table. Judging by the questions they posed, they had the most open, non-judgmental minds in the group, hence the cleanest slates to project Hugo's thoughts. Similarly, in Ancient Greece, peasant girls were trained to be oracles for the priest class of the religion of Apollo.

There were doubters among the group, such as François-Victor, Hugo's younger, more literate son, who was translating Shakespeare's complete works into French. While the tables rarely worked for the Englishman Pinson, the object of Adele's obsession and her undoing, he did get a brief message from 'Byron'. Hugo himself was often in doubt. In a left-hemispheric manner, he insisted on names, dates and first-class proof of the existence of the spirits; but his right-hemispheric interest, conscious or not, never waned. The spirits' immense respect for him and the 'sublime' quality of their messages were too compelling to reject. Hugo was principally concerned that the spirits were copying *his* work and passing it off as their own. He repeatedly commented, 'Did you know that I've already published verses very similar to that? I've thought the same thing myself for many years. I've used almost the exact same metaphor already'.

All in all, the collaborative atmosphere, with some participants who had suffered childhood trauma, others who had suffered loss of loved ones, and the whole experience of being exiled in an inhospitable, bleak, weather-beaten environment played their role in the séance phenomenon. Jaynes's 'collective cognitive imperative' explains how this might have occurred on an island where people already believed in the presence of ghosts. Among them: a decapitated man who 'wandered' during the full moon; *la Dame blanche*, who had committed infanticide; *la Dame noire*, an ancient druidess who had burned her father on a dolmen in a ceremony; and *la Dame grise* of unknown origin. In his memoir written nine years later, Vacquerie, a great enthusiast at the time of the séances, later denied what he saw and heard himself. Back in Paris, away from a world where the wind, waves and island ghosts reigned, he transformed into a doubting Thomas. In his own book, *William Shakespeare*, Hugo said the experience of the talking tables merited scientific observation, but divorced himself from the idea of direct divine inspiration:

> *Donc écartons le trépied. La poésie est propre au poète. Soyons respectueux devant le possible, dont nul ne sait la limite, soyons attentifs et sérieux devant l'extra-humain, d'ou nous sortons et qui nous attend; mais ne diminuons point les grands travailleurs terrestres par des hypothèses de collaborations mystérieuses qui ne sont point nécessaires, laissons au cerveau ce qui est au cerveau, et constatons que l'œuvre des génies est du surhumain sortant de l'homme* (Hugo 1864/2003: 78)

> [Let us separate ourselves from the three-legged table. Poetry belongs to the poet. Let us be respectful toward the possible, about which no one knows the limits, let's be attentive and serious before the extra-human, from which we come and what awaits us: but let us not diminish the great terrestrial workers with these hypotheses of mysterious collaborations that are not at all necessary, leave to the brain what belongs to the brain, and let us remark that the work of geniuses is from the superhuman coming from man. (Translation mine)]

Even with Vacquerie's first-person account, we will never know the entirety of the séances. According to Jean Gaudon, who has written most extensively on the subject, three of the red notebooks in which they were recorded have long since disappeared and some sessions were purposely deleted. However, the fact that Hugo did not even have to be in the room for 'contact' to occur argues for a group dissociative phenomenon.[2] Given that the words produced so closely resembled his own poetry, some sort of telepathic communication with Hugo cannot be dismissed, especially given an instance where only he had foreknowledge of the expressions the spirits used. Sometimes he left the table early and the dictation continued without missing a beat. Either some participant was taking dictation from Hugo's mind at a distance or another had a dissociative ability to construct a

---

[2]   'Métempsychose' was in fashion between 1850 and 1860. Even in Flaubert's *Madame Bovary* (1856), the lover, Rodolphe, claimed *'attractions irrésistibles'* were because of some previous life (Grillet 1929: 13).

Hugolian-style message through his or her close affinity to him and his work.

The paused séance experience brings to mind Jeremy Taylor's poem from the old Greek woman in Chapter 3. Just as Taylor's intense research and writing on matriarchal myth primed and organized his dissociative poetic overflow, Hugo's politico-social preoccupations — revolution, democracy, European unity and equal rights for women and children — and personal preoccupations — the senseless loss of innocent life — took center stage. When asking questions, Hugo focused on God, the cosmos, reincarnation and the authenticity of the spirits' voices. The spirits told him to ask his questions in poetic form, compelled him to write and to ask for infinitely more:

> *Moi, si j'étais à ta place, je demanderais tout ou rien; j'exigerais l'immensité, je ferais des sommations à l'infini, je lèverai ma barricade jusqu'au dernier étage du ciel, je ferais une révolution complète, je voudrais tout savoir, tout tenir, tout prendre...*

> [If I were in your place, I would ask for all or nothing; I would demand the immense, I would plead before infinity, I would raise my barricade up to the final stage of heaven, I would make a complete revolution, I would want to know all, hold all, take all... (Translation mine)]

The spirits informed Hugo that he was to found a new religion based on their revelations about reincarnation and that the Book of the Tables would be the Bible of the future. In addition to Shakespeare, the daunting and diverse list of those who 'visited' the tables included: Racine, Molière, Diderot, Rousseau, Marat, Charlotte Corday, Robespierre, Chateaubriand, Hannibal, Machiavelli, Aeschylus, Socrates, Plato, Aristotle, Dante, Byron, Walter Scott, Galileo, Moses, Jacob, Joshua, Isaiah, Luther, Louis XVI, Mohammed, Jesus Christ and the local ghosts. Despite *their* greatness, the 'spirits' invariably emphasized *Hugo's* and insisted he get to work: '*Au travail, grand homme!* [To work, great man!]'.

But other séance participants got personal responses as well. Vacquerie received a workable outline and a title for a text he was writing. Charles got a name for a heroine in a novel he was writing and the title 'The Winds of the Tomb' for the Book of the Tables. Vacquerie asked Shakespeare to dictate a new play and the bard obliged. The guillotined French poet, André Chenier, completed a poem he had been writing before his execution and described his death in language sounding very much like a modern NDE report. Hugo asked Rousseau if women would ever have equal rights. Rousseau, in keeping with his writings, replied that women were like children and their fundamental right was to be a mother. When Hugo's daughter Adèle asked what happens to women who do not have children, Rousseau condescendingly replied that she was a flower and should leave the proceedings, go into the pure air and let him discuss with the thinkers (18 December 1853, in Guadon & Gaudon 1968: 1262–3).[3]

It was rather dispiriting for Mme Hugo that the men got long, involved responses (her husband had just received twelve rhyming couplets from an ancient Greek poet), and she, the strongest believer, got short, dismissive responses. Did the spirits even know it was her birthday? In response to her pleading tone, Machiavelli broke through to tell her that 'heaven is misunderstood by most men and dead souls surround her like angels' wings, gently singing her name and drinking her bitter tears'. Women are like 'churches for the dead' (14 December 1853: 1259–60). As her mind seemed to develop in the séance process, she eventually questioned certain inconsistencies in spirits' messages, and came to the conclusion that the whole experience had been designed to help her husband accept the possibility of contacting the dead, especially their daughter. Mme Hugo was one of the first to recognize the tables' remarks as highly reminiscent of Hugo's own theorizing over the past twenty years. On another occasion, she asked why Shakespeare was stumbling while composing. He replied: 'The human spirit is a cage. By imprisoning myself within you, I must submit to the laws of the prison, I work… I become the human poet, maybe greater but no longer free. I create laboriously and feel on my forehead drops of sweat, those tears of human labour' (27 January 1854: 1292).

When the séances moved out of their Marine Terrace home to another house for several months, the table talk continued as before, but without Hugo's participation. He was careful to be absent when 'Shakespeare' started dictating a play that resembled too closely a play he had been writing, showing an anxiety of influence. On 27 December 1853, Hugo returned mid-séance. Reading the transcript of the session up to that point, he learned that Balaam's Ass had been revealing an idea that exactly mirrored one of his own. In April 1854, another animal spirit, the Lion of Androcles, asked permission to use part of a line of Hugo's poetry that was only known to the two of them. In a note appended to the transcript, Hugo replied '*Oui*', going immediately to his office to confirm that he had indeed written the same *hémistiche*[4] in March, referring to flowers as 'nocturnal coquettes' (l. 20, 'Vénus', *Toute la lyre*). The metaphor could only have come from his own mind, since no one else had read it.

Many of the 'spirits' communications seemed random. When would Hugo's unwritten plays be finished and performed? *In two years.* When would the United States of Europe be established? *In six years.* How long would Bonaparte reign? *Two years.* As in Hélène Smith's transcripts, many

---

3   The following references, which I have translated from the séance proceedings, refer to their '*Présentation*' in Hugo's *Oeuvres completes*.

4   A *hémistiche* is half of the 12-syllable alexandrine.

of the dialogs started as a suggestive question, followed by an inter-pretative prodding, met with a *'oui'* or a *'non'*. Even minimalist answers reflected the Hugolian world. In response to Vacquerie's questions, 'Civilization' responded that speech was its greatest force; verse was the greatest form of speech; and to reach the greatest heights, one must suffer. Asked who had contributed most to realizing civilization on Earth, the personified 'Civilization' replied 'Hugo', then Chateaubriand (Hugo's adored precursor), then Napoléon (whom Hugo greatly admired). Hugo was exhorted to finish his novel: *'Grand homme, termine les Misérables* [Great man, finish *Les Misérables*]'.

The séance messages curiously contained the spirit of twentieth-century New Age teachings and spiritualist beliefs. Divine oneness solves all prob-lems. God is all alone everywhere; God is the great tear of the infinite. (Hugo himself had previously written almost the same line of verse in his poetry: *'Dieu, larme de l'infini'*.) Metaphors are indeed keys to the otherwise ineffable kingdom of the One. The following French quotes are from Mutigny (1981), with page numbers noted, followed by my translations:

1. All is One in the universe: *'L'unité est le total de Dieu. Il n'y a pas de chiffre mille; il n'y a pas de chiffre cent; il n'y a pas de chiffre dix; il n'y a pas de chiffre deux; Dieu ne compte que jusqu'à un'* (40); *'Dieu est partout et nulle part'* (47). [Unity is the total of God. There is no number thousand; there is no number one hundred; there is no number ten; there is no number two; God only counts to one. God is everywhere and nowhere.]

2. All you need is love: *'Amour, amour, tu es la solution suprême, tu es le dernier chiffre.'* (41). [Love, love, you are the supreme solution, you are the last number.]

3. Being is multiple and each part contains the whole: *'L'homme n'est pas un moi simple, c'est un moi complexe, dans son épiderme, il y a des millions d'êtres qui sont des millions d'âmes'* (42). [Man is not a simple self, he is a complex self, in his skin, there are millions of beings who are millions of souls.]

4. 'Idea' dictates dietetic laws that could be considered quite modern: abstention from eating many meats and fish, especially shellfish; an accent on eating fruits and vegetables; and the general edict to eat only animals and plants one has personally raised or cultivated. In a very un-French way, 'Idea' also prohibited eating *foie gras*!

5. Hugo had been chosen to receive the divine Word, elements of which would be widely diffused after his death. He was a new messiah. Jesus himself spoke to Hugo to complete the teachings that would guide the further evolution of humankind. Those who believed unfailingly in religious dogma paled before *'le grand interlocuteur de Dieu'* —the genius, poet or prophet who actually speaks with God, i.e. Hugo himself, who had

already used the same metaphor in his poetry (15 March 1855, in Mutigny 1981: 54).

6. Death is not to be feared; it will be a true cosmic liberation: *'La Mort c'est la vie affranchie et superbe'* (l. 49, in Mutigny 1981: 56). [Death is life superbly set free.] As dictated by Shakespeare—in French!

In a hyperbolic message, 'Joshua' described a delirious, universal system: there is no Dantesque underworld; rather the brain of Man confines *within itself* an infinite number of unknown and imperceptible suffering souls. Like a great nest, each fibre of the brain contains a thinking soul. *Man alone* imprisons millions within him, all tortured for their sins in previous existences. In this scheme, which predated David Bohm and Karl Pribram's holotropic mind/universe by a century, the simplest soul contains an exact copy of all the souls of the known universe—including men, animals, plants, stones and stars—as they, in turn, contain all others in infinite regression.[5] 'Joshua' says that since everything is contained in everything else, if the entire universe, save one grain of sand, were to perish, God would smile and toss it into space shouting: *'Sortez million de mondes!* [Exit, millions of worlds!]' (28 December 1854: 1450).[6] In a new Big Bang, everything would begin again as in the Hindu cosmic conception.

In 1831, Hugo had already described his own overpopulated mind in a poem called *'La Pente de la rêverie'* (The Daydream's Slope). In this poem, the poet looks out his window and sees all of his artistic friends, both living and dead, swelling the imaginary scene, at first pleasantly, then with nightmarish proportions. Eventually, he sees an immense pile of humanity, from all time and all space, a Babel toiling together uncomprehendingly. Then, everything disappears; nothing remains but a limitless ocean. The poet's spirit dives in and gaspingly arises, having found eternity. The solitary dip into boundlessness quells his mounting angst. The poem ends in rebirth from a world of pain into a shimmering, naked new existence.

---

5   As Michael Talbot (1991) said in *The Holographic Universe*: 'The idea that consciousness and life (and indeed all things) are ensembles enfolded throughout the universe has an equally dazzling flip side. Just as every portion of a hologram contains the image of the whole, every portion of the universe enfolds the whole. This means that if we knew how to access it we could find the Andromeda galaxy in the thumbnail of our left hand. We could also find Cleopatra meeting Caesar for the first time, for in principle the whole past and implications for the whole future are also enfolded in each small region of space and time. Every cell in our body enfolds the entire cosmos. So does every raindrop, and every dust mote, which gives new meaning to William Blake's famous poem [quoted previously]' (50).

6   Alternatively, Hugo may have been creating imagery based on the seventeenth-century German philosopher Leibnitz's theory that individual 'monads' make up the universe and each contains a reflection of the whole universe.

Reincarnation is not only a consolation for loss of the body, but also an explanation for bad karma in this life stemming from sins in a former one. Rebirth also allows a return to another, possibly better, mother. The cyclical round of life is a matriarchal concept versus the patriarchal notion of earning your way into one eternal existence. Mystical 'oneness' is hypothesized to restore a sensation of fetal oneness with the mother. Freud called this the 'oceanic feeling' and the sea itself is a maternal symbol.[7] In New Age thought, we choose our new mother in our next incarnation.

Hugo's mother fixation seemingly knew no bounds. It manifested in 'Shakespeare's' play dictated through the tables: 'The enormous God of the abysses, the tempests and the winds, becomes the loving father of his child who shows himself to be so tender, so devoted, so gentle, that men say to him: my mother! Creation pulls back her claws and only caresses remain' (translation mine). God, for Hugo, was a Mother! 'Idea' said to Mme Hugo: 'A woman's voice, and especially that of a mother, is the earthly music that we hear the best. Motherhood is a nest, and we are the birds [*La voix de la femme, et surtout de la mère, est la musique terrestre que nous entendons le mieux. La maternité est un nid, nous sommes les oiseaux*]'. In his poem, 'Le poëte est un monde enfermé dans un homme', written in 1854, Hugo named all the great poets, comparing the famous personae growing within their skulls to fruit within a mother's womb. First, God was a Woman, now poetic creativity is a maternal flowering. Nature speaks to poets, catching them on a thorny bramble if a dangerous path is ahead. Her voice in the wind murmurs vaguely as they pass: there go Shakespeare and Macbeth, Dante and Béatrice, Molière and Don Juan. She withdraws her thorns to allow the great minds who speak with 'phantoms' to pass. Hugo remained faithful to a life-long support of the rights of women and children (see Sarde 1983: 123–5). When 'Death' came to the tables, bemoaning the loss of Aeschylus, Dante, Shakespeare and Molière, he used a metaphor dredged from Hugo's unpublished repertoire: the great poets' tombstones are like maternal breasts that humanity comes to suckle in vain.

What the French Said about Hugo's Séances

Jean de Mutigny

French doctor Jean de Mutigny analyzed in detail the transcripts from the table. He believed that the phenomenon was similar to automatic hand-

---

[7] Bachelard (1942) says that water cradles us, puts us to sleep and gives us back our mother. It is also an invitation to die: death in water is a special and most maternal death. Yet, as the sun dips into the sea, so shall it be reborn from the depths. But, on a less metaphoric note, Hugo's daughter would die a horrific, watery death in 1843.

writing (*écriture automatique*), which the French Surrealists would also explore to unleash the powers of the unconscious. Mutigny emphasized that the messages never surpassed the intellectual level of the table's participants and that hidden memories could be brought back by the stress of the séance itself. Under all circumstances, the messages retained a whiff of the nineteenth century about them, no matter which epoch the spirits had come from or what their profession had been. They all sounded like Hugo.

Mutigny's diagnosis was dissociative identity disorder, or '*dédoublement de la personnalité*'. In France, only two personalities are recognized, and that is what you typically find in the psychological literature. Believing Hugo incapable of trickery and rejecting the possibility of unconscious communication between father and mediumistic son, Mutigny claimed that Hugo's 'second' personality took over to write up the manuscripts, unconsciously expressing his own thoughts about cosmic forces and immortality. For Mutigny, Hugo's enormous imagination and intuition, his hallucinatory experiences, his persecution complex, his plays on words and neologisms were all evidence of a late-onset mental disorder similar to schizophrenia, called *paraphrénie fantastique*. In this disorder, a cosmic belief system could exist separately from everyday reality. Hugo, he said, was a delusional megalomaniac for saying he was a messiah whom Moses, Mohammed and Jesus had contacted. Yet, softening his diagnosis, Mutigny concluded that poetic inspiration *depends* on genius splitting. By exploring a space and time outside of reality, madness makes art timeless, immaterial, magical and even of divine origin and Hugo was the greatest poet.

Charles Baudouin

Psychoanalyst, philosopher and poet Charles Baudouin attributed a number of psychological complexes to explain Hugo's poetry and séance materials. First of all, he diagnosed a 'forehead' complex. Hugo did possess a large forehead, which he took to signify his superior intelligence. Even God, poetized, had an enormous and mysterious forehead, according to Hugo. Baudouin interpreted Hugo's complex as overcompensation for being the youngest of three brothers. Secondly, Baudouin noted that great artists often suffer some serious problem in their relations with their parents, citing basically the same biographic materials as I do, but attributing the problem to an Oedipus complex rather than a compromised sense of self from early maternal deprivation. Baudouin claimed that Hugo rejected his real father, considered a tyrant for separating him from his mother, in favor of an 'ideal father' image, a combination of Chateaubriand and Hugo's godfather, Lahorie, who was his mother's lover. Mother, Nature and Poetry, then, became entwined, as a sublimation of his Oedipus

complex and flight from his father. Baudouin saw Hugo's frequent allusions to a cosmic abyss as a refuge in the maternal womb, not a *lack*
needing constant replenishment.

All of these predisposing complexes culminated in an unreal, visionary,
twilight state, with depersonalization and derealization, while exiled on
Jersey, which became a breeding ground for poetic genius and a grandiose
identification with St. John of Patmos. *'Ramener Tout à l'Un, et sentir en soi la
présence de l'Un, c'est n'être pas éloigné de se considérer soi-même comme le
centre unique du Tout* [To conflate the All in One, and to feel in oneself the
presence of the One cannot be far from considering oneself as the centre of
All That Is' (Baudouin 1943: 129–30 [translation mine]). In a verse from *Les
Feuilles d'Automne*, Hugo does, in fact, admit he thought the heavens were
illuminated just for him. Baudouin, like Mutigny, forgave the visionary his
hubris.[8]

Jean Gaudon

While recognizing Baudouin's ability to *read* Hugo, Gaudon deplored the
overly deterministic psychoanalytic approach to his writing. With his considerable talent for literary sleuthing, Gaudon focused on the question of
authorship, paying attention to the dates of Hugo's published poems and
the dates of the séances. Which came first, his verses or the table's pro-
nouncements? Gaudon erroneously stated that Hugo did not attend the
first séances, but the intense emotion surrounding his dead daughter's
purported appearance certainly drew him in. Hugo said his son Charles's
excessive 'animal magnetism'[9] multiplied his intelligence five times to produce the table's seemingly supernatural revelations.[10] Gaudon theorized
that the humiliated lesser son became his all-powerful father's guide to the
great mysteries, without having an inkling of how he did it.

As Gaudon said, Hugo eventually conflated all of the spirit voices into
one source, the *Bouche d'ombre*, the Shadow's Mouth, thus bypassing the
question of their self-proclaimed identities. Indeed, what the table
expressed unconsciously as a long enumeration of nocturnal animal cries,
became a long, consciously created poem entitled *'Ce que dit la bouche
d'ombre'* (What the Shadow's Mouth Says [13 October 1854]). Here, Hugo
translated into rhyming couplets of exquisite beauty what the séance

---

8 Twentieth-century French poet Paul Claudel considered Hugo as much a
   visionary as William Blake (in Baudouin 1943: 132).
9 A term Franz Mesmer introduced in the previous century. Mesmer claimed
   that certain people had 'animal magnetism', a special fluid with curative
   and paranormal powers.
10 James Merrill makes note of this in his *Changing Light at Sandover*, but
   ascribes the power to Hugo rather than to his son, Charles.

transcript had actually said. In this poem, we find the fullest account of the Hugolian universe: *'Tout parle'* (Everything speaks); *'Tout est plein d'âmes'* (Everything is full of souls). *'Le mal, c'est la matière'* (Matter is evil) (Hugo 1856/1972: ll. 12, 48, 82: 485-7). On a ladder of incredible lightness of being, material weight indicates relative distance from God: diaphanous angels and archangels are close; man, animals, plants and rocks increasingly far. A specter that guides the poet on a tour of the sentient world, like Dante's Virgil underground, said we build our own prison in the next life with our actions in this one. The murderer becomes his own victim, nailing himself to the heaviest matter. Specters sometimes watch souls leaving the bodies of the living, quickly snatched and imprisoned in a wolf, a flower or a stone. Hugo was convinced the tables only confirmed what he already knew or completed what he had only half-expressed. In fact, there was an imbroglio of conscious creation, unconscious expression and conscious amplification (see Gaudon & Gaudon 1968: 1173-6). When Hugo insisted that the spirits explain their great mysteries more fully, they refused. The mystery *must* remain — a convenient excuse, one might say, camouflaging his inability to answer some of his own questions.

Gaudon claimed that Gustave Simon, publisher of the first record of the séance proceedings, purposely deleted sessions that questioned Hugo's creative precedence over the tables. But, since the generative flow was certainly coming from Hugo's own mind, unconsciously, telepathically or as memes passed on to his collaborating friends and family anyway, what does it matter? Hugo's head, no longer wobbling on his infantile neck, was now supported by worshipful friends, family and a coterie of ghosts who reminded him of his greatness and urged, even demanded, that he write.

In this shadow world exiled on foreign soil, Hugo could express poetically an extraordinary interplay of analytical questioning and dissociative, collaborative creativity. He worked on two major collections of poems, *Les Châtiments* and *Les Contemplations*; two long fragments, *Dieu* and *La Fin de Satan*; as well as his famous novel, *Les Misérables*. Before then, Gaudon said, Hugo wrote when he had time, with brief moments of fecundity followed by a poetic desert. When it came time to publish, he threw everything he had together, along with more pieces to fill out the volume with an imposed order, and wrote his preface. Leaving behind the sporadic and the linear, Hugo's creative process now became an oceanic *'magma poétique'*, a limitless proliferation, an 'abundance of inspiration', a 'dynamism of imagination' (Gaudon 1969: 201-8). Hugo's discourse with the dead solidified his internal division, liberating and authorizing him to become truly *himself*.

Gaudon dismissed the idea of thought transmission in the table phenomenon. Rather, the table's revelations crystallized, confirmed and systematized Hugo's intuitions on reincarnation and propelled his creativity, while allowing him to subordinate, but not forget, the pain of his daughter's death. But how do we explain the *mechanism* of this dissociative creation? I would again propose that pre-existing dissociative tendencies from childhood traumas erupted full-force in the exile environment. In 1837, he had already published a collection of poetry called *Voix intérieures* (Inner Voices) with a persona named Olympio, with whom he conversed as his own double.

During exile, an enormous sense of loss, both political and personal, in tandem with group collaboration, *amplified* his internal divisions giving them full rein to speak, albeit in one voice. The boy who could write poetry in his sleep joined the man who spontaneously created poetry from the sound of horses' hooves. What seemed like a sensational collaboration with the unknown was the tapping out of poetic speech with the help of his family and friends on ideas long percolating in his own brain. As the tenor changed and the tone became increasingly bombastic, commanding and imperialistic, like the voice of Yaweh, Hugo's thoughts were emboldened beyond his conscious self's. Yet, he still followed the trajectory established early on, i.e. a universal system where the dead are trapped in animate or inanimate material commensurate with the crimes they committed in pre-vious lives.[11] The *Bouche d'ombre* demanded that both the trapped and the material that traps them, the nails of Christ's cross for instance, suffer and deserve to be extolled in verse.

Hugo, with his losses, terrors and genius, was capable of gaining new ground through unconscious thought processes, just as other great minds have solved problems in their sleep or in moments of relaxation. The exile itself was like one long sleep designed to solve two seemingly insurmount-able problems: the unjust rise of Louis-Napoléon to power and the sense-less death of Léopoldine. The shaman never operates alone: his powers come from the combined belief of the tribe and rituals that call up the spirits. Hugo's coterie of followers gave him an energetic stimulus, whether through attention and intention, constant conversation and reading his works, or through right-hemispheric mind-to-mind contact with intimate others suffering their own exile and losses, along with a group sensitivity to powerful forces of nature. Since everyone in the group revolved like satellites around the life and work of Victor Hugo, it is not surprising that the tables' talk resembled his own so closely.

The whole experience was a means whereby an already right-enhanced mind, now magnified by present sorrowful emotions, held together a group of exiles seeking entertainment, knowledge and comfort in a foreign, weather-beaten terrain. Using metaphoric language, visual analogies,

---

11   Hugo's reading influences could have furnished his ideas on reincarnation, the ladder of beings and the theory of stellar migrations (Gaudon 1969: 236).

nature symbolism and authoritative Presences — poets, prophets, scientists, political figures, Biblical characters and literary abstractions — all dissociative offshoots of Hugo himself — generated a Big picture structure to contain the whole universe, explain its workings, and exalt the Poet who spoke for them. His inner circle evoked the spirits who talked in the dissociative voice of Hugo, *even when he was not present*, to explain what he already knew, push his concepts further and tell him what he needed to hear in order to create his masterpieces. Claudius Grillet's (1929) analysis was closest to mine, accepting that telepathy was the most likely explanation, but still not explaining how it could happen. He called on scholars and theologians to explain the phenomenon. The transformation of the traumatized mind to communicate through right-hemisphere-to-right-hemisphere non-local consciousness may be that explanation. A return to emotional homeostasis and ultimately survival are very strong impulses, not to be denied, to those who suffer most.

The dialogs with the dead ended abruptly on the heels of an incident where a sometimes participant, Jules Allix, descended into madness. Dissociating can have lethal effects on lesser minds than Hugo's. The entire group would be expelled from Jersey to Guernsey when a French exile's inflammatory article thought to impugn the Queen of England was published. Following nineteen years in exile, Hugo returned to France with the fall of the Empire in 1870. He had the greatest intentions for political reform, but disillusionment, followed by enormous sorrow in witnessing the bloody events of the Commune, brought a return to poetry. At his deathbed, Hugo said the family angels, Homer, Jesus Christ, Dante and Shakespeare were all present (Robb 1997/1998: 522). A crowd of hundreds of thousands of real admirers followed the pauper's coffin he had demanded to its stately repose in the Pantheon, along with the greatest minds of France.

Conclusion

Moving from myth to mediumship, what Keats *wrote* about in his 'Hyperion' poems, Hugo lived in his island exile. The persona Hyperion and the real Hugo both stewed in the right-brain cauldron of negative emotion and specters haunted them. Voices reminded both of their greatness and commanded them to act. Poetic gods and mythic goddesses were forced to humanize their sayings to be understood in the land of the living. The poetic voices of both Keats and Hugo bred mythic mothers to replace their familial losses. Wind and waves informed divinity in 'Hyperion'. They also 'spoke' directly to Hugo. The right hemisphere, with its love for embodied being, stands poised to interpret all that is broad, contextual, environmental and novel, seizing consciousness to relay a message, sometimes with metaphors so harsh they hurt. The mythic gods of the past must give way to the New. We can no longer suckle dead tombstones.

Chapter Six

Dictating Others and Surrogate Mothers: Rainer Maria Rilke and W.B. Yeats

Rainer Maria Rilke

Rose, oh pure contradiction, desire,
To be no one's sleep under so many Lids.
— Rilke's self-authored epitaph

Epiphany and the Terrifying Angel

Standing on a bridge in Toledo, Spain, Rainer Maria Rilke caught sight of a meteor falling beautifully through the night sky and *right through him*. Interestingly, his visit to Spain had been set in motion when the spirit of an 'unknown woman' had beckoned him from this bridge (Freedman 1996: 350). Rilke had been involved in a series of séances in the fall of 1912 at the instigation of his patron, Princess Marie von Thurn und Taxis, and had served as recording secretary. He was excited to find answers to his concealed questions through a planchette, a step up in the spirit world from Victor Hugo's table-tapping. Rilke, like Hugo, was part of a culture that valued 'ghosts', and actively sought contact with them as counselors for the creative imagination. The invitation from an unknown woman was compelling, since Rilke had already contemplated going to Toledo to see El Greco's paintings. A mere possibility became an imperative to act, and fortuitous once there.

   Triggered by a serendipitous sighting, Rilke's epiphany seemingly bridged the cosmic and the human, time and space. As we have seen, when mind and nature interrelate to the point of near synchrony, a sensed merger opens the self to a feeling of oneness with the universe. Rilke's further attempts to access answers through a medium or by using a planchette on his own failed. The princess, observing the proceedings only, also claimed she was worthless as a medium; her son, Pascha, like Hugo's son, Charles, assumed that role.

Earlier in the year, the sound of a violent, cliff-side storm converted simmering ideas into poetic speech. In this direct encounter with Nature, Rilke's inspiration came through as the voice of a terrifying Angel who had 'given' him the initial line of his *Duino Elegies*. This dissociative linguistic event resembled Hugo's voice of Nature heard in the wind-blown, wave-crashed environment of his island home. Exile prompted alien voice in both cases: Hugo had spent his long exile on Jersey, then Guernsey; Rilke lived in a perpetual state of self-imposed exile as he crisscrossed Europe, wandered as far as Russia and Egypt, in search of inspiring environments for creative solitude.

Rilke was a voracious reader intent on understanding science as well as his poetic predecessors. He had been reading Hölderlin and Goethe's *Roman Elegies* before his own *Elegies* burst forth (see Gass 1999: 100, 108). Immediately before their uprush, he had been contemplating an important but 'tedious business letter' that needed writing (Sword 1995: 53). Wandering through the windy ramparts of the Duino castle, where he was staying through the generosity of his patron, he stood still to hear an external voice: 'Who, if I cried out, would hear me among the Angels' orders?' (Rilke 2009: ll. 1–2: 283).[1] Arduous prior thought in composing the letter may have prompted his *cri de coeur*, which erupted from his meandering consciousness.

After returning to the castle and completing his letter, Rilke transcribed the entire First Elegy, recalling the angelic figures of Islam, 'and even if one of them pressed me suddenly to his heart: I'd be consumed in his more potent being' (ll. 2–4: 283). As we saw, both Jeremy Taylor and the Hugolian séance goers were able to pause their voices until they tuned in again. Rilke wrote the Second Elegy, and the first lines of the Third and Tenth Elegies, but without the same sense of entrancement (Sword 1995: 56). In a letter to the princess, he described the religious sensation of writing from dictation: 'I am writing like a madman... The voice that now makes use of mine is greater than I; I only rustle like a bush through which the wind has passed, and must simply let it happen' (12 January 1912, in Sword 1995: 54). Writing again four days later, he gushed: 'I must have written with both hands, both to the left and to the right, in order to catch every word granted to me' (in Sword 1995: 54). Metaphoric or not, he may have been expressing a sense that both hemispheres operated at once as the dictation took place. Rilke translator William H. Gass was unromantic about the seemingly dissociative poems: '[He] had written them over and

---

[1]   All other poetry quotes from Rilke will refer to the Snow edition, unless otherwise indicated.

over already. There is scarcely a line, an image, an idea, that we cannot find, slightly rearranged, in earlier work' (Gass 1999: 109). He further elaborated, less eloquently than Rilke: the poems flow because 'the perception will have soaked for a long time in a marinade of mind, in a slather of language, in a history of poetic practice' (Gass 1999: 148). While this description may be true, it does not override Rilke's *sense* of dictation and effortless fluency.

Count C.W., more Elegies and the Sonnets to Orpheus

A ten-year dry period ensued without further visitations, except for the 1920 poems *From the Literary Remains of Count C.W.* Looking for inspiration, Rilke had conjured up an eighteenth-century Count to dictate verses to him. Professor Helen Sword says he was denying responsibility because the poems were not worthy of him (Sword 1995: 66). But, often, voice-inspired writing is processed even though the writer rejects what is being said and it will continue until the text is done. Jung's *Seven Sermons to the Dead* came out that way. Rilke biographer Ralph Freedman eloquently theorized that Rilke's 'fantasy' had 'frozen' the 'fluid logic of dream' into a 'detached language that thrusts itself forward anticipating *The Sonnets to Orpheus*' (Freedman 1996: 462-3).

I would say Count C.W. was more like a waking dream character who, once created, spoke *as though* real, and was not that far removed from Jung's eighteenth-century gentleman, who served as his No. 2 personality in youth; or the Philemon of his adult years, who first came to him in a dream, then conversed with him in the garden.

During three weeks in February 1922, while staying at a loaned château in Muzot, Switzerland, Rilke felt the remaining six Elegies, as well as the entire fifty-five *Sonnets to Orpheus*, burst forth (Sewell 1971: 314-15). Rilke's ability to compose so many sonnets, so quickly, awed Sewell, as she was a sonnet writer herself. She considered Rilke the last of the great 'Orphic' poets of the modern period, following Wordsworth, Coleridge, Shelley, Hugo and Mallarmé (Sewell 1971: 280-2). She considered the *Sonnets* more successful than the *Elegies* because the mythological Orpheus provided 'a narrative framework' to hold Rilke's somewhat bizarre private world in place (Sewell 1971: 215).

What had prompted this belated, supercharged output? For one thing, Rilke had been reading the poetry of Paul Valéry, who described conquering his own twenty-year drought in a poem:

> Patience, patience
> Patience dans l'azur,
> chaque goutte de silence
> est la chance d'un fruit mûr!

[Patience, patience
each drop of silence
is the chance for a ripe fruit!] (cited in Sword 1995: 66)

Rilke later remarked, 'I was alone, I waited, my whole work waited, then one day I read Valéry and I knew that my wait was over'. Here, we see shades of Suzanne Segal's realization, after reading a resonant Buddhist text citing a time period relevant to her own experience. Rilke will describe his inner dictation of the *Sonnets to Orpheus* as 'spontaneous', a 'tempest', a 'divine gift'; he is a 'vibrating vessel' (in Sword 1995: 66-7). Stephen Mitchell, another Rilke translator, waxed mystical about the *Sonnets*: 'These poems were born perfect; hardly a single word needed to be changed. The whole experience seems to have taken place at an archaic level of consciousness, where the poet is literally the god's or Muse's scribe' (Mitchell 1985: 8, cited in Sword 1995: 68).

Death and the Poet

The diary of Wera Oukama Knoop may have been an even more important reading influence. This close friend of Rilke's daughter had wanted to be a dancer, but had died from leukemia at the age of nineteen. When Wera's mother sent Rilke the diary, he replied that the girl's 'ghostly presence' 'commands and impels' the 'cycle as a whole' (Sword 1995: 68). The *Sonnets* were more than a memorial to her; there was a deep identification, and, perhaps, a precognitive intuition of his own demise from leukemia only four years later.

Poet and theorist Edward Hirsch rightly reads most of Rilke's poetry as being 'written in close proximity to death, in striking distance of it... even... to its very last borders' (Hirsch 2002: 40). In Rilke's poem 'Death' (*Der Tod*), written in 1915 in the middle of World War I, the poet described a teacup containing blue death distilled, balanced on a hand, not a saucer. Ghosts around the breakfast table mumble through empty gums, their extracted false teeth show a metaphoric inability to separate completely from this life. The image of the cup on the hand came to Rilke when taking a walk; the ghostly breakfast scene followed after rushing home to capture the first image. The final image, though strikingly different from the rest of the poem, remains familiar for its content: 'O fall of stars / once glimpsed while leaning from a bridge —: Never to forget you. To stay!' (ll. 17-19: 537).

Hirsch saw death in the meteor streaking across the sky and arcing into the poet's eyes. Freedman said it was also the horrible arc of bombs falling in war. Rilke had been conscripted into World War I at age 40 and, for a brief period, had written war poems. The horror of war not only impelled Rilke, but also the entire modernist movement in art and poetry. Yet, the epiphanic moment on the Toledo bridge, an image of beauty, I would say, arose unbidden at poem's end to counterbalance death. Along with the free fall of death, Rilke accented the sense of 'unity' that was 'given' to him, a gift from the gods. The chance splendor in the night sky connecting with his inner being may also represent mother–infant oneness, a joyous weightlessness streaming through space. The recurring image in his poetry always arises with a sense of ecstasy.

Hirsch recognized '*duende*' in a number of famous poets who claimed to receive inner dictation. He acknowledges the 'foreground', without denying Rilke his 'uncanny', 'oracular' compositions:

> There are in truth any number of uncanny moments in Rilke's high lyrics when the mind seems to give way before an incomprehensible mystery and, out of a long foreground, the lines of the poem seem to be forming themselves, as if dictated by a force from without that is also somehow a voice within. They seem visited from the other side. Rilke's receptivity to such oracular moments may be one of his most defining characteristics as a poet. (Hirsch 2002: 40-1)

As we saw with Gass, Sewell and Freedman, literary critics and biographers will likely gauge a poet's first-person accounts by their own experience and beliefs. Sword also accepts Rilke's account. In *Ghostwriting Modernism*, she says, 'Rilke's poetry everywhere bears witness to his mediumistic role of one of the twentieth century's preeminent poets of death' (Sword 2002: 87). In *Engendering Inspiration*, she recognized his poetic inspiration as 'a real phenomenon rather than a mere rhetorical construct' (Sword 1995: 24), not a long-running hoax.[2]

Jungian James Hollis reads Rilke's dictations as the personified splits of the archetypal imagination. According to Hollis, not only poets facing death are privy to the inner daimon, but also each and every one of us; although most lose contact with their 'interior Other' in childhood. It is the task of therapy to rediscover 'the individual yet transpersonal dimension which drives us, and constitutes our linkage to largeness' (Hollis 2000: 40-1).

Rilke himself saw his role clearly, as revealed in a letter: 'Ultimately there is only one poet, that infinite one who makes himself felt, here and there through the ages, in a mind that can surrender to him' (to Nanny Wunderly-Volkart, 29 July 1920, in Gass 1999: 183). In other words, poetic consciousness is collective and collaborative, amassed and interjected since the beginning of time into poetic minds properly receptive to the Primal One. The poet must surrender to the daimon, obey the voice of the Angelic orders, and make visible the invisible.

---

2 She also cites Prescott's 'real ecstatic' inspiration versus a 'secondary' imitative inspiration (Sword 1995: 5).

According to this book's thesis, early trauma predisposes the poet to hear a dissociative Other who can say the unsayable from the vantage of enhanced right-hemispheric processing. By incorporating signs from the outer world into the body's felt sense, a poetics of presence brings meaning through personification, metaphor and symbol. Reading Rilke, one enters the inner sanctum of the self's purest understanding of the universe's greatest mysteries. The poet's sensitivity to Nature and private history intersect in a historical moment of crisis to call forth priestly powers of expression.

Anniversaries are markers for uprushes of memory seeking commemoration. 'Requiem to a Friend' is a long poem Rilke wrote quickly during three days in November 1908, nearly a year after the death of Paula Modersohn-Becker, a talented artist friend. She had left him and Paris to return to her husband in Germany, having no other means of support. There, she died tragically soon after from childbirth complications. Not blaming her husband alone, Rilke implicated all men for depriving women of their worth and relegating them to motherhood, which can kill. Women, he says, like men, must be 'let go' to fulfill their artistic destinies.

As the poet writes, he senses Paula's presence in his room. Although highly praised, she is admonished to return to the dead where she belongs. She sowed the seeds of her own demise. The weight of this world is no place for her ethereal self; shedding the world once and for all should be her goal. But, *perhaps*, from otherworldly distance, she might hear *his* voice and help him overcome the struggle 'between ordinary life and extraordinary work' (Gass 1999: 133).

When the young dancer, Wera, died, a mental commingling might well have occurred when Rilke read her diary and triggered the Sonnets. But as trauma recalls previous trauma, perhaps they were tied at a deeper level to the loss of Paula, *and*, at an even deeper level, to Rilke's Mother complex. Already, in the *Third Elegy*, Rilke spoke to the question of love and loss, unwinding his feelings for the lover and drawing them back to those for his mother, despite his fraught relationship with her and his professed lack of love for her. Yet, it was she who 'started him', and 'over his new eyes… arched / the friendly world, keeping the strange one out. / …[who] stood calmly between him and his surging chaos'. She was the 'night-light' that 'shone like a friend' (ll. 26–30: 297; ll. 1–2: 299). Then again, this may well be a fabricated recollection. Rilke also adorned his dead father with more importance than he had had in life; as did Hugo, who created in his poetry the ideal mother and father he would have liked to have had.

The Mother and the Religious Other

Nonetheless, Hollis the depth psychologist believes Rilke the poet must have intuited the truth about the primal role of the mother: she is the 'imago of the Intimate Other' who lies more profoundly beneath the poet's feeling for his beloved. For everyone, this maternal image, internalized since early childhood, serves unconsciously as the 'template which all other relationships replicate or struggle to transcend' (Hollis 2000: 39). Hollis asserts that 'the power of the mother experience, for men and for women, is, generally speaking, the single greatest psychological influence in our lives'. She is 'mediatrix with the world' and her 'fears, unlived life, and projected desires become part of the internal mythology of the child'. As an adult, the enormity of this early influence remains a 'seductive power' sucking him back to womb-like comfort from which he must escape in order to fully actualize himself (Hollis 2000: 40–1).[3]

But aside from this danger of maternal engulfment lies the more fundamental danger of physical or emotional abuse at the hands of parents. Hollis, the depth psychologist, meets Schore, the neuropsychologist, declaring that abuse and neglect fracture the emergent ego and its sense of self and other. Still, the 'interior other' and the 'Intimate Other' add up to more than the internalized mother for Hollis. Where Hirsch deduces *duende*, Hollis reveals a religious experience. Hollis translates Rilke's 'huge obedience to the spirit' into 'the possession by the daimon who is both personal and universal, terrible and transformative' (Hollis 2000: 36). The Freudian Hirsch sees an electrified excitation, a sexual urge, countering the presence of death. The Jungian Hollis perceives a poet listening in awed humility, his consciousness struck dumb from great suffering as well as the nearness of death. For Hollis, the highly intuitive poet channels 'the will of God' in the oldest tradition of poet as prophet of the divine. He turns absences into presences, transforming invisible mysteries into the metaphors of art. Only those who can process this 'largeness' will fulfill their destinies, poetic or otherwise.[4]

In his introduction to *Poetry, Language, Thought*, Albert Hofstadter explained Heidegger's notion of the poet in similarly religious terms. 'The poet it is who, looking to the sky, sees in its manifestness the self-concealment of the unknown god, bidding the unknown to come to man to help him dwell' (Heidegger 1971: xiv). Poetry is 'a movement away from the thin abstractions of representational thinking and the stratospheric con-

---

[3]   When writing about D.H. Lawrence, Sword noted, 'Speaking in tongues, possessed by his mother's language and vision, the poet perceives living people as ghosts even while his mother, the real ghost, remains vividly and persistently alive' (Sword 2002: 79).

[4]   See also Martz (1998). In Martz's poetics of presence, the voices and visions of Eliot, Pound, Whitman, William Carlos Williams and H.D. give missions, convey warnings, redeem civilizations—expressing the voice of entire nations, whether past, present or future. Their voices encompass this spatial or temporal largeness.

structions of scientific theorizing, and toward the full concreteness, the onefoldness of the manifold, of actual life-experieneee' (Heidegger 1971: xvii). *In neuroscientific terms, poetry is a movement away from left-hemispheric representation of the divided and fragmented real towards a right-hemispheric appreciation of fully concretized reality.* Heidegger himself said, 'The artist remains inconsequential as compared with the work, almost like a passage-way that destroys itself in the creative process for the work to emerge' (Heidegger 1971: 40). *The artist is a sacrificial conduit for the divine.* Respond-ing to Hölderlin's question concerning the role of poets, Heidegger responded, 'To be a poet in a destitute time means: to attend, singing, to the trace of the fugitive gods. This is why the poet in the time of the world's night utters the holy' (Heidegger 1971: 94). *Poetry is a place marker for the gods in times of great crisis.*

In the same vein, Harold Bloom said that Rilke, Shelley, Hugo and Yeats *augmented* their own extraordinary consciousness through admired dead geniuses, but that their ultimate creative source was their 'God within'. Poets must 'ruin the sacred truths'—the title of one of Bloom's books—to create a *new* poetic voice and vision. By surpassing the god-like precursor poet, the younger poet becomes a god himself. But, for Bloom, the strongest poets are those who came before the Cartesian (read *left-hemispheric*) engulfment. The Freudian Bloom says the younger poet wit-nesses 'his Poetic Father's coitus with the Muse', his Mother, with whom *he* now mates, thus usurping the Father (Bloom 1997: 37). The poet is 'daemonized', i.e. split in two, as the 'Holy' rushes in dictating poetry from the stance of the 'Wholly Other' (Bloom 1997: 101).

Coming down from these metaphorical heights, which nonetheless reflect certain basic neurobiological facts of life, I would say that owing to the confluence of genes, early trauma, voracious reading, relational drama and national crises, great poets have the means to call forth the 'gods'—high-order dissociative constructs—to impel their work. Rilke's inborn sensitivity, negative childhood experiences and later losses molded his consciousness toward gender ambiguity, an obsession with death and distance, and a self-professed task to pin down the transience of life with a religious reverence for 'things' brought to life through his gaze. The con-stant tension between solidity and weightlessness, the allure of the absent and the barely there, show these concerns. His animistic appreciation for the plant, animal and mineral world joined him to his predecessor Hugo, who also judged closeness to God in terms of weightlessness. For Hugo, reincarnation into heavy form was the worst-case karma, a punishment for crimes in a previous life. For Rilke, we live one life. The poet's task is to become invisible, while naming absence with enduring words.

Contradictions and the Maternal Wound

Rilke's life stances were highly contradictory. His uncompassionate critic, William Gass (1999), explained: he celebrated childhood, although his own was miserable, and he did not like children; he denied his Czech, German and Austrian selves equally. Merger was his ideal, but distance remained preferable; he relentlessly relinquished love. Yet, Rilke honored his own contradictions. At the poetic level, consciously or unconsciously, he alternately expressed his creativity through worshipful feminine or hard-driving masculine imagery.[5]

Whether following Freud or Jung, we must look for the child behind the man to better understand the psyche and steady drift of Rilke's ethereal poetry. Born weak and premature a year after a sister who died shortly after birth, Rilke was at first closely guarded by a mother who feared losing him — yet was seemingly replacing her lost daughter by dressing him as a girl until he was 5 years old and calling him 'Sophie'. Ralph Freedman (1996) suggested that mother and son played equally at the dress-up game, which bonded them together. The gender transformation also served to placate Rilke's mother when she was angry. As a 7 year-old, he once approached her groomed and dressed as a girl, reportedly saying, 'René is no good. I sent him away. Girls are after all so much nicer' (Freedman 1996: 10).

Nonetheless, it is unfair to blame a child for something he was trained to do. His mother had an enormous influence on him, with her agenda of poetry, religion, femininity, French–language training and refinement from an early age. Freedman says Rilke's mother 'urged poetry upon him before he was even able to read. At seven, he also started copying poems, and knew many of Schiller's lengthy ballads by heart' (Freedman 1996: 10). Actually, all of the penchants Rilke adopted from his mother would serve him well in the cultured world he would come to inhabit.

Gass defined Rilke's childhood in these chilling terms, referring to the lost younger sister: 'Rilke realized that someone else had had to die in order to provide him with a place in life' (Gass 1999: 9). We might add that both van Gogh and Salvador Dalí, also geniuses of invention, began their lives as replacement children. Rilke could not be consistently himself and his mother was equally inconsistent. Cuddled then cast aside, he only managed to attract his mother's attention when dressing as a girl, when ill or when dedicating himself to the poetry and religiosity his mother held dear. Often abandoned to a maid, especially after his parents' separation, he was caught between his father's masculine, military ideals and his mother's preference for the feminine. Sent to a military boarding school at 10 years old, he again felt abandoned; only recurrent bouts of illness

---

5   See Sword (1995) for the best treatment of this poetic gender ambiguity.

brought his mother, his 'saving angel', back to his side (Freedman 1996: 9–20).

Obviously, Rilke had a deep maternal wound. While he claimed his early novel, *The Notebooks of Malte Laurids Brigge* (1910), first translated under the title *Journal of My Other Self*, was not autobiographical, many elements of his own life were in place. Rilke described the mother's desire for a girl and the protagonist's adopting the name 'Sophie' and a female persona to please her. Other personal allusions include a mother obsession; uncanny experiences; accounts of dissociation; voracious reading; laments of loss from women poets; religion; male angels; the notion of being an ear for God; John of Patmos taking dictation with both hands; and the foremost importance of an inner life. Rilke's poetry would go on to extol absence and death, to cry for breathing room—fitting reminders of his loss of self, gender, even his original name—and his complicated relations with women.[6]

His maternal wound would be salved in adulthood by surrogate mothers—Lou Andreas-Salomé and the princess—plus a string of lesser lovers (creative women artists, sculptors, musicians all), even a neglected wife and daughter. Tormented from childhood by a mother alternately overbearing and absent, Rilke would adopt the same *modus operandi* with women: to court and to abandon. He was a man who constantly craved women to redeem, mold and affirm him, yet needed to let them go and be let go just as much. Being pinned down by a woman precluded the great poetic task; yet, to gain her love, he pried her heart in writing, whether poetic or epistolary. Seducing both self and other through words, he could overcome his early undoing. Affirming the androgynous self could even be construed as womb envy. In *Letters to a Young Poet*, Rilke wrote: 'And in the man, too, there is a motherhood... physical and mental; his engendering is also a kind of birthing when he creates out of his innermost fullness' (cited in Freedman 1996: 193).

We saw how Paglia implicated the Great Mother's rape of the feminized male in her appraisal of Keats's 'Hyperion' poems. Sword too equated a 'poet's self-feminization' with 'rape by the inspiring Other' (Sword 1995: 57–8). Dissociative theory configures it differently: the abused child internalizes the offender, whose traits he or she takes on (see Howell 2005). Men whose mothers were properly present need not write dissociative poetry with its longing, distance, and still, receptive stance.

Gass (1999) reminds us how the mythic Orpheus was torn apart by women. Hence Rilke the poet praises them when young and dead, while condemning women in general because they can become mothers. Yet, it is

---

[6]   For an in-depth treatment of Rilke's conflicted gender identity and attitude towards sexual relationships, see Dorothee Ostmeier (2000).

not motherhood itself he condemns, but life outside the womb, where Rilke the child *had been* 'tenderly tethered' ('The Eighth Elegy' in Gass 1999: 174), then 'tentatively touched' ('The Second Elegy' in Gass 1999: 194). Subsequently, he would be cut loose as a girl doll, a puppet, a military lad. Rilke the striving poet became awash in distant wanderings and relationships. While we all must be cut loose, Rilke's rupture seemed infinitely more debilitating. Like the cracked cup in '*Der Tod*', his infant mind was later compared to a newborn bat's jagged trace against the evening sky in 'The Eighth Elegy'.

Gass said Rilke's Angel was a projection of the poet's better self. He did not recognize conflictual mothering in its origin. I believe this primal splitting allowed the poet's right hemisphere to dictate to the left in the analogical language of Nature, bringing solace and making commands, as if from a good Mother, but in the more authoritative voice of a male Angel. Judging by Hugo and now Rilke, a male poet has difficulty 'gently outgrow[ing] the breasts of a mother' ('The First Elegy' in Gass 1999: 192).

Having lost the steady affection of his own mother, who outlived him, and too preoccupied with his work and self-protective distance from women, Rilke would fail as a father too. He put little effort into raising his only daughter. She, on the other hand, remained dedicated to him, even after his death, collecting his letters and publishing his works. His paternal failure may have been her demise. Sadly, she committed suicide in 1972.

William Butler Yeats

Cast a cold eye, on life on death. Horseman pass by.
— Yeats's epitaph, from 'Under Ben Bulben'

Epiphanies and Anomalies

It may be no coincidence that Rilke was considered the best German-language poet of the twentieth century and Yeats the best English-language poet. Early childhood trauma, exile, anomalous experiences, occult influences and practices—a full right-hemispheric regime—punctuated both life stories. Although contemporaneous with Rilke, the older Yeats would outlive him by thirteen years. Rilke had never read the poetry of Yeats, but the Irish poet was familiar with his. In fact, he wrote the famous lines that would become engraved on his tombstone after reading an essay on Rilke (Maddox 1999: 373).[7] Both drew strength from the voices of the dead. While the obsessed and leery Rilke would push back their presences to the underworld realm, Yeats actively sought their counsel. Learning about his pre-

---

[7]   Saddlemyer (2002) cited a letter by George Yeats saying her husband had written the poem as 'a direct result of reading that essay "Rilke and the conception of death" by William Rose' (Saddlemyer 2002: 618).

vious existences and expecting successive renewals on the Great Wheel, Yeats could indeed 'cast a cold eye, on life on death'.

Yeats described what sounded like an epiphanic experience around the same time as Rilke:

> My fiftieth year had come and gone,
> I sat, a solitary man,
> In a crowded London shop,
> An open book and empty cup
> On the marble table-top.
>
> While on the shop and street I gazed
> My body of a sudden blazed;
> And twenty minutes more or less
> It seemed, so great my happiness,
> That I was blessed and could bless. ('Vacillation, IV')

The 'open book' would suggest a reading-inspired moment saturated with significance. Yeats's wife told Richard Ellman that it had only happened once, yet the scholar found another instance in 'Stream and Sun at Glendalough'. Here, a Yeatsian 'sunstroke' resembles Rilke's 'starstroke':

> What motion of the sun or stream
> Or eyelid shot the gleam
> That pierced my body through?
> What made me live like those that seem
> Self-born, born anew?

Without knowing the neuroscience of these altered states, or probably having experienced one, Ellmann (1948/1979) beautifully described their effect as 'accesses of feeling' where 'two worlds converged, offering totality of being, insight into the heart of things, foreknowledge of what was to come' (Ellmann 1948/1979: xviii).

Whereas Rilke had participated in only four séances and received his dissociative poetry on his own, Yeats would need the close collaboration of his wife to make contact with spirits on an almost daily basis for three years. The greater the trauma, the more overt the dissociation will be; the closer the collaboration, the greater the creative output will be. Yeats had not been raised as a girl for five years; rather, he would marry a very intelligent young artist with uncommon mediumistic powers.

But Yeats did not rely on his wife alone. Raised in an Irish home where fairies, banshees, omens and tall tales were common fare, Yeats was no newcomer to the occult. In his *Reveries over Childhood and Youth*, he mentioned his anomalous sound and sight perceptions, prophetic dreams and the beginnings of his psychical research. While still in Ireland, he attended a séance where strange things occurred: 'a drawer full of books leaped out of a table when no one was touching it, a picture had moved upon the wall'. Even stranger were his own physical reactions: his shoulders and hands twitched; his whole body was then 'thrown backward on the wall'. When the spiritualist mediating the event moved towards him, Yeats struggled vainly against a force so compelling and violent that his movements broke the table.[8] [9] Terrified, but unable to remember a prayer, he recited the opening lines from Milton's *Paradise Lost*. The experience put him off séances and table-tapping for years, leaving him to question whether the 'violent impulse' had come from within or without him (Yeats 1916: 126–9).[10]

In 1912, around the same time as Rilke's *Duino Elegies*, Yeats would make contact with a tamer spirit named Leo Africanus through an American medium in London. On a later occasion, Yeats would mention 'Leo' to a medium who spontaneously channeled the 'spirit's' voice. 'Leo' gave him an easier method for making contact: Yeats could write letters to him, and he would respond using the poet's own hand. Neale Donald Walsch (1995) would use the same method in his late twentieth-century *Conversations with God*. Walsch explains here what happened in a fit of frustration:

> This time, rather than another letter to another person I imagined to be victimizing me, I thought I'd go straight to the source; straight to the greatest victimizer of them all. I decided to write a letter to God… To my surprise, as I scribbled out the last of my bitter, unanswerable questions and prepared to toss my pen aside, my hand remained poised over the paper, as if held there by some invisible source. Abruptly, the pen began moving on its own. I had no idea what I was about to write, but an idea seemed to be coming, so I decided to flow with it. (Walsch 1995: 1)

No mysticism, no séance, no trance state, just a friendly conversation. I tried a method, recommended in a Jungian seminar, where 'I' wrote to 'me' in a journal dedicated to the conversation. The 'other' me, speaking from its detached perspective, brought surprising revelations. I would argue that people with increased right-hemispheric dominance, from whatever cause, might more easily generate their 'higher' authority in automatic responses. In a recent study, a team of researchers found that an individual's 'higher-thinking' abilities were significantly correlated with the amount of gray matter in the *right* anterior prefrontal cortex and the structure of neighboring white matter (Nauert 2010: n.p.).

---

[8] Ellman (1979/1948) said Yeats banged his head on the table.

[9] Yeats (1937/1973) described a similar incident that occurred when he was about to tell his wife a story about a Russian mystic: 'a flash of light fell between us and a chair or table was violently struck', as though to keep him from telling the story (Yeats 1937/1973: 15).

[10] A so-called 'catalytic exteriorization phenomenon' occurred during a vexing conversation between Jung and Freud. A bookcase next to them made a 'loud report' without being touched — twice (Jung 1961/1989: 155).

Yeats was thrilled with the idea that his inner psychological conflict might be 'an external battle between a living man and a dead one, between this world and the next' (Ellman 1979: 199). His inner 'Leo' would explain the extroverted mask he sometimes wore versus his normally shy, cautious self, as well as the incredible power he only felt when writing verse, presumably as Leo (Ellman 1979: 199–200). As with Hugo, Yeats suspected the information coming from the 'Other' side, including things he had been thinking about for years, was rooted in his own imagination. 'Leo' insisted on his separate existence and told Yeats to look him up in an encyclopedia (Longenbach 1988: 190). Here, Yeats discovered that Leo Africanus was a sixteenth-century Moorish poet who had written a treatise on Africa in Italian, after falling into the hands of Pope Leo X. But, while the sixteenth-century and the twentieth-century poets were conjoined in one mind, Leo affirmed, 'I am your opposite, your antithesis because I am in all things furthest from your intellect & your will' (Longenbach 1988: 192).

A Marriage of Minds

Yeats's major dissociative adventure began after his marriage at 51 to the 25 year-old Georgie Hyde Lees. Yeats had been in love for years with Maud Gonne, a fiery Irish woman who did not share his passion, and, more recently, with her beautiful daughter, Iseult, who was twenty-two months younger than Georgie. Fighting for her marriage, Georgie at first faked a symbolic response meant to settle her new husband's mind on his final choice for a marriage partner. However, to her own amazement, she felt 'her hand grasped and driven irresistibly', as told to Ellman by Georgie Yeats herself, scrawling out barely legible messages (Ellman 1979/1948: xiii). Yeats (1937) spoke of another curious incident where, after worrying about something he had said to a Japanese dinner guest, he heard a loud, clear voice declare: 'You have said what we wanted to have said' (Yeats 1937/1973: 17). Whereas only he had heard it, his wife reportedly wrote the same sentence, totally out of context, in a letter she had been in the process of writing at the other end of the room. Such a bizarre phenomenon suggests an auditory hallucination, but, at the same time, a strong telepathic connection between the married mystical aspirants.

Enthralled by her talent and ready access to the Other side, Yeats nonetheless insisted on reading the same books Georgie had read prior to their marriage to verify the novelty of her automatisms. When certain that neither she nor he had been influenced by their readings, he wrote 'The Gift of Harun Al-Rashid', a long wedding poem about an Arabian court poet who marries a woman who can 'speak mysteries':

> I heard her voice 'Turn that I may expound
> What's bowed your shoulder and made pale your cheek';
> And saw her sitting upright on the bed;

> Or was it she that spoke or some great Djinn?
> I say that a Djinn spoke. A live-long hour
> She seemed the learned man and I the child;
> Truths without father came, truths that no book
> Of all the uncounted books that I have read,
> Nor thought out of her mind or mine begot,
> Self-born, high-born, and solitary truths,
> Those terrible implacable straight lines
> Drawn through the wandering vegetative dream… (Yeats 1925/2008: 101)

Interestingly, both Rilke and Yeats chose an Arabic mystical connection, more esoteric and distant than a Christian one, to frame their poems.

Atypically Lateralized Lovers

Just how did this newlywed speak mysteries from the marriage bed? Or, more aptly put, why was such an intelligent young woman so prone to dissociation? First of all, she was a left-handed, female artist (Saddlemyer 2002), hence more open to presences and claims of the paranormal, according to both Persinger and Radin. From childhood, she had been a sleep-walker (a mixed state of consciousness), suffered migraines and had a history of poltergeist and ghostly contacts 'announcing themselves through strange noises, aromas, and cryptic messages' (Maddox 1999: 80). Poltergeist activity, as Jung theorized, arises in adolescents whose burgeoning sexual feelings require an outlet. Secondly, it was a consensual and possibly genetic affair. Her mother, Nelly, 'was interested in occult matters; she read cards and believed in psychic powers' (Saddlemyer 2002: 27). The family considered both Georgie and her brother Harold to be clairvoyant.

A general factor in mediumship, which Jawer and Micozzi (2009) consistently found in their survey, was a history of trauma and loss, shown to increase right-hemispheric activity over the less emotional left (Anthes 2010: 9). Georgie had learned by telegram of her adored, but alcoholic, father's death when she was 17.[11] Eight years later, she was still mourning him and 'the very sight of a telegram gave her a shock' (Maddox 1999: 60). Death surrounded her. At the time of her early séances, not only her father but also four uncles had died, either in the war or from illness (Saddlemyer 2002: 57). Her spirit communicators would later identify the loss of her father as her first 'moment of crisis' (Saddlemyer 2002: 121). This loss may explain, in part, both her psychic ability and her marriage to a great man the same age as her father. She also claimed a voice had directed her to Yeats on a London street (like Rilke's female spirit to the Toledo bridge), which was a 'chance' encounter that revived Yeats's interest in proposing to her (Maddox 1999: 64).

---

11   Her father, grandfather and great-grandfather had all died of alcoholism (see Saddlemyer 2009: 243).

Georgie's obsessive interest in the occult, through séances and voracious reading, as well as her membership in the Order of the Golden Dawn, indicate a dramatic search for esoteric knowledge and contact with the dead. Fleck *et al.* (2008) found that people (like Georgie) with atypical language lateralization—dominant language function in the right rather than the left hemisphere or spread out bilaterally over both hemispheres—show increased paranormal effects. Testing which hemisphere makes lexical decisions, Fleck's study found that those who score high on magical thinking use both hemispheres equally, while those who score low, the less magically inclined, use only the left hemisphere. Nor can the genetic factor be denied. Making meaningful associations between loosely connected words or ideas is typical of the right hemisphere and both schizophrenics and creative thinkers share this trait (see Leonhard and Brugger 1998; Lindell 2014). Relatives of schizophrenics can be especially creative and '[a]t least one of Georgie's uncles and two of her cousins' were schizophrenics (Saddlemyer 2009: 126). High emotionality certainly played a role in her dissociative abilities as well: she often burst into tears when coming out of a trance (Saddlemyer 2009: 108).

What about Yeats? His handedness sheds some clues about his brain organization. He parted his hair on the right like a left-hander most often does. An online sketch attributed to his artist father shows Yeats reading a newspaper in his left hand. Other photos show him waving a cigarette held in his left hand or extending his left hand out to a bookcase. In several poses, he sits left hand atop right or with his left hand under his chin. Even the illustrated cover of the 1937 edition of *A Vision* showed him holding flowers in his left hand, extended to a scornful woman with her back to him.

While the photos indicating left-handedness are not definitive, Yeats's inability to learn to read until around 9 years old and his life-long problems with handwriting and spelling all suggest dyslexia. Functional MRI studies on right-handers show that dyslexic readers activate regions in the right hemisphere, whereas non-impaired readers use only the left.[12] Ambidextrous children, with mixed cerebral dominance, are more prone to language problems. In dyslexia, the left hemisphere area responsible for reading is not functioning properly, causing difficulty in decoding the phonemes of a written word and using incorrect graphemes when spelling. Yeats, for instance, regularly wrote 'sleap' for 'sleep'.

When doing automatic handwriting, Georgie would sometimes use mirror or even upside-down writing. Both Harper and Saddlemyer suggest it was so that 'the medium' herself would not understand things she wrote that might be troubling to her. But to do mirror writing at all requires a

---

[12] See www.dyslexia.com/science/differentpathways.html.

facility at it. It is a sign of dyslexia and the left-handed Leonardo da Vinci used it most of the time.[13] Since Georgie was an insatiable, precocious reader with a phenomenal ability in foreign languages (not true of dyslexics), the trance state alone may have triggered her mirror writing. Flournoy said that Hélène Smith tended to confuse left and right when entering trance states; perhaps Georgie experienced similar confusion or 'allochiria', the medical term.

Could an eventually voracious reader like Yeats really have been dyslexic, with his left-hemispheric language center compromised? Apparently, yes. First of all, dyslexia is genetic. His father, the artist J.B. Yeats, did not send William to school when he was young to spare him the floggings he himself had experienced as a youth with reading difficulties. Rather, he took on the job of teaching his son himself, after various family members on the maternal side had failed. 'Angry and impatient', he once threw a book at his son's head trying to tame his 'wandering mind' (Yeats 1916: 24). While heavy-handed, he did teach his son by reading aloud and lecturing. The father's early intervention may have helped compensate for his son's learning disability. A long-term fMRI study says that dyslexics who learn to avoid the dysfunctional left-hemispheric region in favor of more forward regions in the left and the right hemispheres *can* become good readers. Those who continue to rely on an improperly functioning left hemisphere will remain poor readers (Shaywitz and Shaywitz 2005).

Still, Yeats never wrote dissociative, flawless poetry *à la* Rilke; rather, he retooled his words endlessly to get them right. 'A line will take us hours maybe', Yeats complained (Qtd. in Sword 1995: 1). Reportedly, he started with rhymes, then filled in the rest (Maddox 1999: 87). This poetic approach may appear to contradict the right-hemispheric theory, since rhyming has been shown to be a left-hemispheric phonological function. A recent Welsh/Australian linguistic study confirmed that males use the left hemisphere exclusively when asked to identify similar rhymes, whereas females can use both hemispheres (Lindell and Jarrad 2007). However, all of the subjects in this linguistic study were strongly right-handed, so this might not apply to Yeats, especially if he had to transfer reading functions to the right hemisphere. We might speculate that his initial left-hemispheric reading problems produced compensatory right-hemispheric language dominance, allowing him to rhyme with that hemisphere as well. I recently learned for the first time in our long friendship that my friend who channels 'angels' is dyslexic too.

---

[13] Fritjof Capra (2007) said the great artist's spelling and syntax were highly idiosyncratic; he strung words together, used abbreviations, shorthand and no punctuation. But it was also 'highly eloquent, often witty, and at times movingly beautiful and poetic' (Capra 2007: 168).

A Mathematical Corollary

In *A Vision* (1925), Yeats would join theoretical forces with French mathematician and physicist Henri Poincaré.[14] After stating Poincaré's theory that 'time and space [were] the work of our ancestors', Yeats asserted dramatically: 'With the system in my bones I must declare that those ancestors still live and that time and space would vanish if they closed their eyes' (Yeats 1925: 128). Poincaré (1897) was actually saying that we inherited our concept of space from our ancestors, who had developed them in relation to their own bodies. For Yeats, however, the ancestral became spiritual and, unless the dead lived on in a dimension where time and space were relative, we would not be able to contact them. Either it is so, and the spirits are real; or it is not, and we are fabricating their voices from our own or our collaborator's recombinant memories, linguistic fluency and intelligence. At least in public, Yeats would eventually tout the latter theory, but while the 'miracle' was happening, he was a believer.

Even the scientific Poincaré's special mental powers fit our model of the atypically organized brain. Ambidextrous, absentminded, with poor eyesight, he possessed a highly developed internal visuospatial genius on par with Einstein's, who has also been called dyslexic and did not speak before he was 3 years old. At 5 years old, Poincaré was ill with diphtheria and had a paralyzed larynx. His early linguistic trauma impelled him towards isolated reading with a visual capability to instantly locate the exact page and line of everything he had read. In his teens, he developed the ability to solve complicated math problems simply in his head while pacing around (West 1997: 135–7). Later, while on a leisurely geological excursion he made a major mathematical discovery. Stepping onto a bus, he suddenly intuited that 'the transformations I had used to define the Fuchsian functions were identical with non-Euclidean geometry' (Poincaré 1913). As we saw, Jaynes referred to this 'dark leap into huge discovery' to show how preparation through conscious forethought, followed by incubation in a relaxed state, precedes the 'light bulb over the head' effect (Jaynes 1976: 44).

Sex, Philosophy and Mysticism

Yeats wrote the first edition of *A Vision* as if everything he said were literally true; yet all was based on Georgie's automatic scripts. Never mentioning her, he attributed the work to the fictitious characters that had dialogued with each other. A number of critics, including those who have

---

[14] Poincaré's theories loomed large in the early twentieth century where discoveries in science stoked new movements in the visual arts and poetry, from Cubism to Surrealism to Modernism. Spiritualism itself was considered a scientific undertaking.

closely studied the original handwritten papers, claim that Georgie was, in fact, the major force behind them. The scripts served to hold the marriage together by directing the couple's sex life and orchestrating the birth of their children. Ellman, though, said that Yeats's personality dominated everything (much like Hugo's in his séances). Certainly, since Yeats was formulating the questions, he would have influenced the answers, even if, as the 'spirits' claimed, *they* were guiding his questions. But we should keep in mind Georgie's dissociative abilities and the possibility of right-hemisphere to right-hemisphere communication in the couple.

Before marrying, Yeats had closely collaborated with the American poet Ezra Pound in the isolated Stone Cottage in Sussex, during three cold winters from 1913-1916. At this time, Yeats complained that 'he had no child, "nothing but a book" to offer his ancestors' (Longenbach 1988: 8). During his stay at Stone Cottage, where he wrote his letters to Leo Africanus, Yeats also practiced 'intellectual vision', a contemplative method accessing visionary experience, comparable to Jung's active imagination method or Epstein's waking dream (Longenbach 1988: 50). Through his readings, Yeats further developed the theory of the *Anima Mundi*, soul of the world and repository of all memories, any of which could be brought to consciousness through the intercession of angelic spirits (Longenbach 1988: 224-5). Passionate and violent moments from the historic past were said to recur again and again—an idea that would resurface in the scripts and play an important role in *A Vision*. Research during this intensely collaborative period with Pound laid the groundwork for the occult writing sessions begun with Yeats's new wife.

Yeats and Georgie were both mystically minded and voracious readers with active imaginations. Saddlemyer (2002) demonstrated that Georgie's extensive readings in the occult prior to their marriage did indeed resurface in the scripts.[15] In some ways, Georgie seemed to be directing the scripts: to make her husband stick to poetry; to emphasize her importance in the process; and to spare herself the lengthy, daily sessions at her husband's insistence. 'Ameritus' declared that Yeats 'must write poetry' and give up the scripts. The 'spirits' constantly reminded Yeats that the success of the psychic adventure 'depends on the love of the medium for you—all intensity comes from that' (cited in Saddlemyer 2002: 221). The

[15] Yeats never went to university, feeling unable to pass the entrance exam; yet he read voraciously on his own: 'Jung, Freud, Nietzsche, Pythagoras, Plotinus, Swedenborg, Buddha', and, of course, Blake (see Maddox: 5). Actually, it is hard to imagine that he read everything she had read. She had read William James, Henri Bergson, Hegel, Descartes, Pico de la Mirandello, Ptolemy, Plotinus, Holderlin, Blavatsky and Swedenborg. She was fluent in French, German, Italian and Spanish and read all the great works of the esoteric tradition in the original (see Saddlemyer 2002: 4).

'spirits', constantly on the alert for Georgie's well-being, expressed concern when she was fatigued. When Georgie grew weary of the automatic writing sessions, her control, 'Thomas', suggested a trance method of speaking while asleep. Via the 'spirit' instructors, Georgie also attempted to limit her husband's outside visits to mediums, while swearing him to secrecy about their own sources.

Yeats was more sexually reluctant, but eagerly sought to hone a philosophical system. The sexual, promoted by the young, newly married Georgie, constantly battled the philosophical for expression in their joint séances. By the time of the second edition of *A Vision*, Yeats would declare that '[s]exual love becomes the most important event in life, for the opposite sex is nature chosen and fated. Personality seeks personality. Every emotion begins to be related to every other as musical notes are related. It is as though we touched a musical string that set other strings vibrating' (Yeats 1937/1973: 88). Yet, as the couple sought answers to their personal problems, the 'spirit' replies would have increasingly historic and cosmic significance, along with anomalous external manifestations.

Combining right-hemispheric geometry and the desire for a child, a grand 'system' arose around the idea of waxing and waning two-thousand-year cycles of civilizations. Interlocking cones or gyres, 'the fundamental symbol of [their] instructors' (Yeats 1937/1973: 68), barely camouflaged the conjoined male and female sexual principle. At the same time, the all-purpose interlocking cones depicted the self and the antithetical self, as well as the objective outer world of reason and the inner world of emotion, desire and imagination (an apt breakdown of left- and right-hemispheric differences).

The resolution of opposites became the ultimate goal of life, leading to Unity of Being, much like Jung's individuation process, but it also reflected the lovers' merging minds. The couple began to have complementary dreams with images 'that complete one another' (Yeats 1925: 140), as Jung did with Toni Wolff. The mediumistic hand of Georgie would provide significant images and phrases for Yeats's creative process. For example, Georgie 'first wrote that the "soul of world [is] in centre" of the historical gyres, and that the modern age is in chaos because "the world's civilization is apart from the centres"' (Harper 2006: 212). In the hand of the poet, this insight would become the wrenching opening lines of 'The Second Coming', as apt now as they were in 1920: 'Turning and turning in the widening gyre / The falcon cannot hear the falconer; / Things fall apart; the centre cannot hold; / Mere anarchy is loosed upon the world'. The instructors had come, after all, to give Yeats metaphors for poetry.

Reading and Reincarnation

Regular sexual activity was prescribed to produce a child; not just any child, but the reincarnation of the dead child of an aristocratic ancestor on George's side, as her husband now called her. Love seeped out from the tomb to relate to the living, as it had in the reincarnation fantasies of the two Helenes (see Chapter 4). The ancestor 'Ann Hyde', speaking from the Other side, saw *herself* as the medium and Yeats as *her* husband (Maddox 1999: 107). The Yeatses felt 'chosen' to produce a kind of messiah within the bosom of their spiritualist religion. A bit less grandiose than Hugo's communicators who had named *him* the new messiah, the Yeatses' coming baby would fill the role. As Yeats badgered the communicators for ever-more specific questions about the 'system', George received instructions from 'Thomas' 'to read up on the fifth, eighth, ninth, sixteenth, and seventeenth centuries' (Maddox 1999: 123). While a tall order, George complied and more precise messages followed. The pregnant George was growing irritable, as were her 'communicators'. On 26 February 1919, the awaited child arrived: a girl, to be named 'Anne', like her forbearer, reincarnated before the son.

The 'spirits' affirmed that people are 'born again and again', in familial or loving relationships, as in mother and son (like Ann Hyde), husband and wife or brother and sister. A community of affiliated 'spirits' comes to exist so that acts of 'cruelty and deceit' committed in previous lives could be expiated in later ones. The tyrant becoming the victim, the victim becoming the tyrant is a pattern still found in past life stories, as seen in Chapter 4. Relationship roles would shift until the individual became purged of wrongdoing. An individual's life would then become so good, that one was conscious of perpetual good luck and said to be harmonized (Maddox 1999: 192). In certain cycles, one could choose the body in which to be born. One telling scenario appeared to explain Maud Gonne's stubborn rejection of Yeats's long love for her: a man's 'wilful refusal of expression' in one life would be repeated in another, when a woman would refuse him (Maddox 1999: 199–200).

As a hierarchy of daimons, guides and ancestral beings assembled, they reasonably suggested that we the living are the source of *their* knowledge, but that the dead have ready access to *our* deepest thoughts across time and space. A Record exists 'of all those things which have been seen but have not been noticed or accepted by the intellect' (Yeats 1925: 184), an apt description of cryptomnesia. All 'images, languages, forms of every kind used in communication from spirits, have passed through living minds whether in the past or in the present' (Yeats 1925: 206). They deliver what is already known, whether in dreams or through a partner, like George, who makes her mind hollow to receive the information.

Moving through twenty-eight phases of the moon, human personality was said to develop in successive incarnations, as seen in New Age and Eastern religions. Yeats's spirits ranked his poetic predecessors, as Hugo's had done, assigning the former to the same phase as the much admired Dante and Shelley. Yeats read broadly in history, looking for commonalities with the emerging system. In the 1937 edition of *A Vision* he excitedly referred to German historian Oswald Spengler's *The Decline of the West*, not only because it confirmed his own intuitions and the messages from his 'communicators' (see Callan 1975), but also because he had used 'whole metaphors and symbols that had seemed [Yeats's] work alone' (Yeats 1937: 18), again reminding us of Hugo's experience. Yeats even suspected that he and Spengler somehow communicated telepathically or tapped into the '"unmeasured mind" of the past' (Callan 1975: 595).

Yeats had heeded the communicators' instructions not to read philosophy while receiving their transmissions; however, once the proofs were in for *A Vision*, his readings convinced him that nothing could have inspired the system 'except the vortex of Empedocles' (Yeats 1937: 20). In a letter to Mrs. Shakespear, dated 23 June 1927, Yeats said, 'I write verse and read Hegel and the more I read I am but the more convinced that those invisible persons knew all'. In a letter from 28 December 1928, he wrote again to Mrs. Shakespear, saying 'George's ghosts have educated me' (in Ellman 1979: 266). Although George's psychic abilities were greater, the collaborative, synchronizing energy of both seekers resulted in an enormous trove of dissociative creativity.

The American Tour

Traveling by train on an extended tour of US cities, the couple engaged only occasionally in the scripts, but supplemented them with Tarot card readings. Occasionally, they sensed anomalous smells or heard external voices. In California, the scripts become 'sleaps'. George could fall asleep instantly, proffering otherworldly wisdom that the poet wrote down in his notebooks. When husband and wife meditated separately, they claimed synchronous visions. Yeats's memoirs often speak of more archetypal synchronicities in his dreams, visions and chance readings (see Yeats 1922/ 2010). 'Spirits' were also said to be able to 'consult books, records, of all kinds, once they be brought before the eyes or even perhaps to the attention of the living' (Yeats 1925: 188).

Maddox (1999) tells how George, passing by a bookstore in Chicago, heard a voice say, 'Fourth shelf from window third from floor, seventh book'. She located the book, which turned out to be, coincidentally or not, Freud's *Totem and Taboo*, which contained the answer to a recurring mental image that had been puzzling her. Uncanny messages revealing book references are not uncommon. Jung had a waking dream in which a dead friend appeared to him and lead him to his house. Standing on a stool, the man showed Jung the 'second of five books with red bindings which stood on the second shelf from the top'. The next day, Jung actually went to the dead friend's house, finding a translation of Emile Zola's works. Volume 2, in that exact location, was called *The Legacy of the Dead* (Jung 1961/1989: 312–13). Jung found the title, if not the content, significant. In French, *Le voeu de la morte* actually means *A Dead Woman's Wish*.

Continuing on the US tour, Yeats received a ceremonial sword from the Japanese consul in Portland, Oregon. The poet heard a loud external voice declare: 'quite right that is what I wanted'. Earlier in the day, George had penned in upside-down writing, 'sword = birth', 'FISH = conception'. The script, the gift and the voice seemed to confirm the coming of another child. Indeed, the awaited son, Michael, arrived on 22 August 1921. The following summer, the Yeatses declared they would give up all forms of mediumship. George, though, would slip back on occasion, sometimes involuntarily. No messiah, Michael eventually became a senator.

Parenting Poetry

How did Yeats become such a famous poet while addicted to the occult? Ellman saw the poet's entire stance in life as a revolt against a loquacious, domineering, skeptical father. The man was indeed angry and violent, used fear and humiliation to teach his son, and was intrusive and opinionated when he went to school. Whereas the father had rejected religion and become an artist, the son found solace on the opposing path, veering early into a world of dreams, myth and religion. But, as we saw, the father also had a positive influence on his son, reading poetry aloud to him, 'continually for six or seven years' and impressing upon him the superiority of dramatic poetry over all forms of literature. Yeats senior lectured to his son and wrote him artistic and intellectual letters. He also brought friends with similar interests to the family home, despite his wife's dis-ease with them (Ellman 1979: 24, 28). Unlike most fathers today, John Butler Yeats advised his son 'never to think about the future or of any practical result' (Yeats 1916: 106) and to 'avoid regular employment because of its threat to creativity' (Maddox 1999: 204). Yeats's siblings would all be drawn to the creative arts: his brother, an artist; his sisters, artists and book publishers; his daughter, Anne, an artist too, like her mother.

But the role of Yeats's mother might be equally, if not more, important. Maddox called her a 'silent, undemonstrative, expressionless' depressive and said categorically that the 'SECRET OF YEATS [was] that his mother did not love him' (Maddox 1999: 189–90). Indeed, a depressed mother plunges into negative emotion as the prefrontal region of the right hemisphere dominates an underactive left (McGilchrist 2009: 64). This negative emotional organization will be reinforced in the infant's mind, leading to a shy, withdrawn adult.

Stressed or depressed mothers are 'more likely to give birth to children who become mixed-handed' (Anthes 2010: 7). Children with bilateral hemispheric organization tend towards schizotypal personality traits (Asai *et al.* 2009). In a study of 250 students, researchers found that the higher the ambilaterality of their subjects' hand preferences, the greater their magical ideation (in Kaplan 2006: 10, citing Barnett and Corballis 2002). If we add the genetic predisposition to dyslexia on the father's side to the environmental effect of prenatal and early childhood maternal depression, Yeats's dissociative tendencies, introversion and magical thinking were practically assured.

Early lack of emotional connection to the mother was followed by loss and physical dislocation. As Maddox tells the story, when a preferred younger brother, Robert, died suddenly, the 7 year-old Yeats happily set his toy boat at half-mast, but his grieving mother became all the more absent. The youngest Yeats child died only ten months after her birth in London. The depressed mother and the younger children would remain in Sligo after their summer holiday; Yeats, the oldest, would be sent back to England alone with his father. In his *Reveries*, Yeats mentioned his separation from his mother and siblings, stating that 'whatever [he] had most cared for had been taken away' in his move to England (Yeats 1916: 35).

Yeats would long for the elusive maternal figure, although his memory of her remained dim (Yeats 1916: 31). He remembered clearly, though, that his mother had never read a book nor cared at all about painting. She only spoke with animation to her servant, a fisherman's wife, as recounting tales about their people. After his mother suffered two seizures, Yeats transformed absent memories into fiction. As she lapsed into total emotional unavailability, he reconstituted her in the novella *John Sherman*, where the central character 'is an only child, a young man who lives in a town exactly like Sligo with his devoted widowed mother' (Maddox 1999: 198). With father and siblings lopped off in one fell swoop, this veiled, wish-fulfilling autobiography reminds us of Rilke's similar strategy in the *Notebooks of Malte Laurids Brigge.*

Exiled in London, a shy, awkward, Irish boy, with no talent for studies, was provoked to uncontrolled rage attacks in his English school. When older, Yeats described his 'harassed' school days in a tellingly dissociative metaphor: he started fighting his adversaries 'without any intention… as if [he] had been a doll moved by a string' (Yeats 1916: 39, 35). In a dissociative precedent before moving to England, Yeats had wondered why he did not possess a 'voice of conscience', then heard one suddenly whispering in his ear. As he later wrote, 'From that day the voice has come to me at moments of crisis, but now it is a voice in my head that is sudden and startling. It doesn't tell me what to do, but often reproves me' (Yeats 1916: 9). Socrates' daemon worked in the same manner.

Doubly damned and blessed, Yeats turned to 'psychical research and mystical philosophy' to escape his father's influence (Yeats 1916: 105) and kept an occult diary of his anomalous experiences, including 'sleeping, half-waking, and day-dreaming visions'. Strongly identifying with his predecessor poet Blake, Yeats reported a vision of 'two books full of pictures of marvellous and curious beauty', one of which 'contained lost poems of Blake and the other... doctrine... that had influenced him' (Ellman 1979: 128). Magic, he claimed, had made his vision possible. To reinforce their affinity, Yeats tried to make an Irishman of Blake through the paternal line, all the while insisting that Ireland alone in Europe could be a holy land.

Attracted to the fiery and beautiful Maud Gonne, the poet was reconstituting his unattainable mother complex. Told he first met her in 1542 when they were lovers, he understood she was from a lower class and the devotion had been one-sided. In the eighteenth century, George was reportedly in love with a man not her husband and she had hurt someone deeply, requiring three incarnations to expiate that wrongdoing. This eighteenth-century husband was a 'very beautiful man', a counterpoise to Maud and her beautiful daughter, Iseult, who, in turn, now 'overshadowed' Yeats's relationship to George (Harper 2006: 203). The entire 'system' he co-constructed with his wife seemed an elaborate rationalization of his obsession with Maud as much as his wife's machinations to supplant her and produce children. This triangular love affair, we may recall, is where the whole dissociative adventure had begun.

Conclusion

Yeats's life story shows him buffeted on either side by early parental influences — the depressive, absent mother and a domineering father. Clearly, he had his father's intellect, but identified with a mother who had been steeped in magic, Irish folklore and the souls of the dead. Early paranormal experiences, resulting from atypical brain lateralization and traumatic experiences, conditioned his mind to understand the warring sides of his personality through occult practices and voracious reading. Marriage to a learned, artistic woman with strong dissociative tendencies, under conflicted emotional circumstances, engendered a life rich in mysticism, poetry, prose and drama, along with political service and a true chance at love.

Yeats (like Hugo) repudiated his mediumistic adventures in subsequent writings: 'Some will ask whether I believe in the actual existence of my circuits of sun and moon... to such a question I can but answer that if sometimes, overwhelmed by miracle as all men must be when in the midst of it, I have taken such periods literally, my reason has soon recovered'

(Qtd. in Harper 2006: 244). George, on the other hand, 'was never comfortable positing that the communicators were only secondary personalities or the products of her own memories' (Harper 2006: 165). She had always shied away from admitting her role in the scripts, keeping her husband's poetry in the forefront.

Interestingly, opposite sex daimons had expressed themselves — in Yeats a female, in George a male (Harper 2002: 308). Inner androgyny seemed to broaden the self. Similarly, reincarnation theories posit alternate gender embodiment is successive lives. In a paper written in the early twentieth century, Dr. Anita M. Mühl (1929) detailed her amazement in the 'latent talents' of her subjects: 'writing poetry or stories; composing music; illustrating and designing and map drawing' (167). One subject perfectly reproduced a drawing of a foot showing every metatarsal, but was totally unable to do so consciously. Others spoke fluently in foreign languages long forgotten. Ouija boards were a springboard to automatic writing, which improved facility.

Ambidexterity could also manifest in the dissociative state. Subject Violet Z, from a family of three generations of automatic writers, stopped the practice after marrying. Later, however, Ms. Z was able to write when Dr. Mühl, actually a friend, placed the pencil in her *left* hand. Writing automatically with her non-dominant hand, the right then wrote as well, but as a *male*-gendered spirit, with the two fighting each other for dominance in a written conversation (Mühl 1929: 178–9). Atypical lateralization helps explain this case as well the extraordinary collaborative experience of Mr. and Mrs. W.B. Yeats.

Chapter Seven

A 20-Year Ouija Board Odyssey: James Merrill and David Jackson

> *Let the mind be, along with countless other*
> *things, a landing strip for sacred visitations.*
> *– James Merrill, A Different Person: A Memoir*

> *There might be some comfort in the recognition of synchrony,*
> *in the information that we all go down together, in the best of*
> *company.*
> *– Lewis Thomas, The Lives of a Cell*

Predecessors, Poets and What the Partners Thought

Merrill began his epic poem, *The Changing Light at Sandover*, with an epigraph from Dante, another visionary poet who had received 'divine' dictation: *'Tu credi 'l vero; ché i minori e ' grandi / di questa vita miran ne lo speglio / in che, prima che pensi, il pensier pandi' Paradiso* XV ('You believe in truth, for the lesser and the great of this life gaze into that mirror in which, before you think, you display your thought' (Harold Bloom translation)). The epigraph signals the importance of the predecessor poet as well as of the mirror, which still stands in the Water Street house in Stonington, Connecticut,[1] where Merrill and Jackson lived and consulted their home-made Ouija board. The 'ghosts' of the board claimed they could see the two of them, their living 'representatives', in the mirror or in any other reflecting surface and vice versa. The unconscious thought that precedes formal thought could be spelled out automatically from deposits in their word banks by recombining memories, information from books read, or borrowed elements from their immediate surroundings. For example,

---

[1]   Per Matthew Zapruder who lived in the house as a poet-in-residence (personal communication).

when Merrill tells Jackson to use an Edna St. Vincent Millay stamp on a letter he was sending, 'Edna' comments on the board.

Merrill once defined this dissociative creative process as getting 'beyond the Self... to reach the "god" within you' (Merrill 1982a: n. pag.). Not surprisingly, Merrill had been reading a number of Jungian books at the time. Mark Bauer (2003), who had access to Merrill's manuscripts and book collection, identified *Four Archetypes, Introduction to a Science of Mythology, The Portable Jung* and *Man and His Symbols*.[2] In addition to mythology, art and poetry, Merrill's other early readings showed his interest in psychology, the occult, religion and philosophy (see Bauer 2003: 195). The autonomous, collaborative, synchronous interrelation between the microcosm and the macrocosm were summed up in Lewis Thomas's book (1974), which Merrill read as well.

The 'spirits' were real, but 'invisible, inconceivable, if they'd never passed though our heads and clothed themselves out of the costume box they found there' (Merrill 1982a: n. pag.). Their fresh perceptions were born of the partners' personal and cultural predispositions, along with the belief, so often encountered in dissociative texts, that existing notions of the 'sacred' must be altered. Merrill admitted that if the self alone were doing the imagining, it was 'much stranger and freer and more farseeing than the one you thought you knew' (Merrill 1982a: n. pag.). Elsewhere, he claimed, in keeping with our dissociative theme: 'the self is extremely ambiguous. There are so many different selves' (Merrill 1982b: n. pag.).[3]

Organization of the Text and Navigation of the Board

The first part of the trilogy, *The Book of Ephraim* (1976), had appeared previously in Merrill's *Divine Comedies*, published the same year as Jaynes's seminal work, and was awarded the Pulitzer Prize in 1977. The second part, *Mirabell: Books of Number* (1978), was divided into sections representing the numbers 0 to 9 on the Ouija board. Reading Jaynes as he was finishing *Mirabell*, Merrill said he 'goggled', because, like the bicameral mind, his poem was 'set by and large in two adjacent rooms: a domed red one where we took down the messages, and a blue one, dominated by an outsize mirror, where we reflected upon them' (Merrill 1979). In *Mirabell*, he said: 'I'm taken in no more than half. The somber / Fact is, I remain, like any atom, / Two-minded. Inklings of autumn / Awaken a deep voice

---

2   *Introduction to a Science of Mythology* is actually by Lévi-Strauss (*Le Cru et le Cuit* in French). Bauer may have been referring to *Essays on a Science of Mythology* by Jung.
3   Thomas (1974) said, 'The whole dear notion of one's own Self — marvelous old free-willed, free-enterprising, autonomous, independent, isolated island of a Self — is a myth' (Thomas 1974: 142).

within the brain's right chamber' (*CLS*: 232–3).[4] Merrill would remain 'two-minded', enthralled by his entertaining 'ghostly' contacts, if somewhat skeptical of their reality. Yet, he needed them, plus some books on science, to write *Mirabell* (Merrill 1979). The third part of the trilogy, *Scripts for the Pageant*, appeared in 1980, followed by the publication of the whole poem as *The Changing Light at Sandover* (*CLS*) in 1982.

To navigate the Ouija board, the partners used an overturned tea cup: Merrill's *left* hand and Jackson's *right* lightly touched it, as the handle swiftly pointed to the successive letters to generate messages. Merrill's job was more physically challenging, moving the cup with his left hand while simultaneously transcribing the words with his right, requiring a measure of ambidexterity.[5] In taking dictation, Merrill was mindful of his poetic predecessors, citing Blake's 'angelic secretariat', Hugo's séances, and, most of all, Yeats's *Vision* papers. George's messages had initially been transcribed in one long scrawl without word breaks, as were Merrill's.[6] Unlike his predecessors, Merrill would convert both questions and answers into poetic form. Some consider Merrill's poetry the best in English in the twentieth century, based on his entire work, especially the lyric poems.

With another drift towards automatisms, McClelland and Slaughter (2007) said Merrill doodled incessantly. In his successive drafts of his poem 'The Doodler', the 'poetic craftsman' revised it fifteen times and the content became more playful in the process (McClelland and Slaughter 2007: 71). As reproduced in *The Paris Review*, the journal pages show form and content, doodling and drafting, intertwined as the poem takes shape. My sense is that left-hemispheric meter and rhyme served to control his right-hemispheric imagination and to keep the poem's tone uplifting.

Similarly, Merrill (1982a) admitted to interviewer J.D. McClatchy that many of the original transcripts of *Ephraim* had been lost, but that 'the main points were copied out over the years into a special notebook'. Enormous and often repetitious, the *Mirabell* materials needed considerable winnow-

---

4   Stephen Yenser (1987) connected Jaynes's theory to the residual effect of the right hemisphere's function of pattern synthesis into a grand design that would be performed by Mirabell's bats. While not explaining how Merrill created the speaking entities, Yenser did say that JM's logical left hemisphere would act as the gods' interpreter, like an early bicameral man's would have done. But, JM, to him, was 'no more Merrill than Dante the pilgrim is Dante the author' (Yenser 1987: 264). Yenser also cited Schiller's idea that poetry must initially suspend logic to let the imagination flow.

5   Alison Lurie (2001) suggested that Merrill's right-handed transcription allowed Jackson to direct the cup.

6   Merrill described '[d]runken lines of capitals lunging across the page, gibberish until they're divided into words and sentences. It depends on the pace. Sometimes the powers take pity on us and slow down' (in Merrill 1982a).

ing. In *Scripts for a Pageant*, the 'design of the book just swept [him] along unaltered' (Merrill 1982a). DJ talked about the strict compulsory nature and time consciousness of this last part of the trilogy:

> They dictated it. All of the Scripts — it was very much a regimen… We had to do it — it started in on this cycle talking about time and the series of moon cycles. We had to get this given amount done in them, and we had to come back at this given moment. They were precise about their schedules, as they were about when the poem would be finished, when it would be published, everything. (Jackson 1979: 35)

Bauer's careful study of the manuscript variations showed that Merrill's 'shaping hand' had been used for clarifying and cutting, but also for enriching with material from new readings. Sometimes, as with Yeats and Hugo, Merrill put his own words into the spirits' mouths. But most of the 'textual variations occur in Merrill's lower-case "commentary", not the "spirits" [small-cap] voices' (Bauer 2003: 199).

What the Spirits Said

Beginning as a parlor game, the partners communicated first with a ware-house fire victim, Simpson, who merely wasted their time until Ephraim showed up.[7] In their daily dialogs, they discovered that Ephraim was a Greek Jew, born in 8 AD at Xanthos. Ephraim had been ordered killed by Tiberius on Capri in 36 AD for the crime of having loved Caligula's nephew. Merrill confirmed the existence of Xanthos on the Asia Minor coast, using his classical dictionary.[8] Xanthos, once the capital of ancient Lycia, is now in modern Turkey. Although not mentioned in the 560-page poem, three times in history — 4292 BC, 540 BC and 42 BC — the city of Xanthos repeated the same horrific tragedy: the men of Xanthos burned their wives and children, gathered against the onslaught of invading foreign forces. Ephraim also alluded to a personal wound: he had come from a broken home, his father enticed from his mother's bed. This was 'the first of several facts to coincide' with Merrill's own life story, including their sexual orientation (*CLS*: 8). Seemingly, a great historic wound and personal suffering were the magnets that had brought Ephraim to comfort JM, as Merrill would be called in Ouija board shorthand.

Seemingly transcending time and space, Ephraim had become Merrill's 'familiar spirit', as Shakespeare before him had spoken of an 'affable

---

[7]   In the poem 'Voices from the Other World', Merrill said the first 'voice' from the board was an engineer from Cologne, dead at 22 of cholera in Cairo. Otto von Thurn und Taxis, whose name we may recall from Rilke's biography, threatened and commanded: 'flee this house' (Merrill 2001: 112–13).

[8]   Place names and dates coincided with the actual reigns of Tiberius and Caligula.

familiar ghost' aiding his rival poet 'nightly'.[9] JM pronounced Ephraim 'Twice as entertaining, twice as wise / As either of its mediums' (*CLS*: 7). As we have seen, close couplings can produce 'Others' with knowledge seemingly beyond the partners' capacities. Were the voices real or multiple minions generated from the raw materials in their commingled minds? Merrill decided to forgo the '[p]lain dull proof' of Ephraim's reality for the 'marvelous nightly pudding' the clever spirit, so much like him, was serving up (*CLS*: 32).

As was the case in Hugo's séances, where the son's hand had to be on the table, then Yeats's wife, who did the automatic writing and 'sleeps', DJ was 'the hand' and Merrill the 'scribe'. In the early days of their odyssey (1955–6), Jackson was found to be highly suggestible and easily hypno- tized. Although a writer too, he was unable to find a publisher for his novel. Jackson gave it up, but 'the untended garden turned to peat, to tar, and eventually fueled our séances at the Ouija board' (Merrill 1993/1994: 77). The emotional DJ reacted 'with tears to messages that had yet to be spelled out'. Further, Jackson had been born with a caul — a piece of the inner membrane that encloses the fetus still attached to his head — a maternal remainder considered lucky and said to confer psychic powers. Jackson was literally a golden boy, whose 'warm, eager tone' attracted everyone, including the 'spirits' (Merrill 1993/1994: 78).

But it was Merrill who had a way with words. According to life-long friend and Pulitzer prize-winning author, Alison Lurie: 'His mind worked faster than that of anyone I'd known: he could answer questions before you finished asking them. Words for him were like brilliant colored toys, and he could build with them the way gifted children build with Lego blocks, constructing and deconstructing elaborate, original architectural shapes and fantastic machines'. She added that Merrill was 'alert to ambiguity and multiple meanings' and gifted in 'poetic, meaningful punning' (Lurie 2001: 10–11). All of these skills are said to be in the domain of the right hemi- sphere. Whereas DJ might have been 'psychic', raking words and imagery from JM's fertile mind, the final poem required honing, revising and tightening into verse, which only JM could do.

Jackson's input was, nonetheless, indisputable: it takes two to work an Ouija board and David most likely helped unleash the flow of words through their collaborative technique. Both men were 'fluent in French, German, Italian, and modern Greek, and Jimmy also knew classical Latin

---

9   Shakespeare, Sonnet LXXXVI. While perhaps only a trope, the 'familiar spirit' might also refer to Shakespeare's inspirational habits, discussed in the next chapter. Bucke (1961/1993) noted that Walt Whitman also claimed to hear a 'Message from the heavens whispering to me even in sleep'. Whitman (2001) also touted his multiplicity to explain his contradictions 'I am large, I contain multitudes' (113).

and Greek' (Lurie 2001: 12). Both played the piano and David was a painter as well. Both were prime candidates for a right-hemispheric adventure with their artistic talents, reading influences, cultural proclivities and privileged lives of leisure.

Judith Moffett (1984) asked us to distinguish Merrill the Poet from Merrill the Scribe like 'two roles as Mind Conscious and Unconscious… as if all of *The Changing Light at Sandover* were an inexhaustibly elaborate dialog between Merrill's waking intelligence and its own unconscious sources of feeling, myth, and dream, with David Jackson as essential catalyst (and supplemental unconscious story-trove)' (Moffett 1984: 161). This is a very fair reading of the dissociative creative process shared between the two men. Moffett alluded to the hemispheric divide, perhaps without knowing it, calling one passage 'an argument between the conscious Poet, who thinks he ought not to depend so heavily on metaphor, and unconscious Scribe, who knows himself unalterably addicted to it'. As she said, '[M]etaphor is in Merrill's marrowbone' (Moffett 1984: 163). Neuroscientifically, we can now say both alliteration and metaphor are processed in the right hemisphere (Kane 2004).

Moffett also alluded to the possibility of telepathy: 'How, for instance, did DJ and JM know that Nabokov was dead — news that reached them first via the Board?' And 'what is it that these two do that others fail to do, which yields such astonishing results?' (Moffett 1984: 174). Moffett said: 'I took the *Sandover* version of that [Nabokov's death] at face value, and Jimmy didn't say anything to contradict it when he read my book in manuscript. But he doctored things a good deal, as we now know, and I just wonder whether they really did hear that Nabokov had died via the board, before finding out by other means' (personal communication, 5 January 2012).

Yet, the nature of the whole Ouija board experience gave her pause for the surprising way that 'Jimmy was actually taking an interest in a big theme of that sort [overpopulation] — he certainly never had before. For a long time I couldn't settle on a perspective from which to view the dictations, but the disconnect between what I knew to be his disdain for public issues, and what was coming through the board, convinced me at least that something very strange was going on' (personal communication, 16 December 2011).

As told to J.D. McClatchy (1979) and related in the poem, Ephraim 'possessed' and spoke through Jackson, in a voice deeper than his own, and stroked JM's face and throat. DJ related seeing Ephraim with long blond hair, resembling 'a certain kind of blond Portuguese you sometimes see in Lisbon' (Materer 2000: 82). Materer confirmed that DJ's notes on this incident were in Merrill's Black Record Book amongst the James Merrill Papers at Washington University in St. Louis, Missouri (Materer 2000: 158, fn. 11). Ephraim was keeping things spicy for JM and DJ.

Could Merrill's homosexuality and his dissociative creativity be connected? LeVay (2011) reported that gay men exhibit greater verbal fluency than straight men, along with less visuospatial ability—a more bilateral brain lateralization similar to women's. Leonard Shlain's (2014) posthumous book on Leonardo da Vinci's brain attributed his subject's genius to childhood trauma, a bilateral brain with a large corpus callosum and hyperconnectivity, *as well as* to his homosexuality, using the same neural argument. McGilchrist concurs that 'abnormal lateralisation' is associated with homosexuality (McGilchrist 2009: 13). I would suggest that right- or bilateral-dominance for language explains Merrill's exceptional facility with poetry as well as his mental collaboration with Jackson.

In Merrill's early poem 'The Will', Ephraim communicated in a mix of French and English, not his native Greek, saying he wanted to possess 'L'OBJET AIME'—JM, the love object—and to warn him of an imminent nuclear disaster. Later, in section H of *The Book of Ephraim*, Merrill described Ephraim's evocation. Using Jackson's notes on the incident, Merrill described Ephraim as a muscled, 'unemasculated Blake nude', aware of how Blake's initial engravings had been later doctored for decency. As 'DJ in trance attempts to lead JM toward a mirror, JM holds back and DJ awakens' (Materer: 2000: 83). Both men feared absorption into the realm of the 'dead' via their amorous ghost. DJ's easy entry into hypnosis and a possession state proves he was the partner with the greater dissociative ability.

Yet, Ephraim's words sounded more like JM, as Hugo's spirits sounded just like him, not his mediumistic son, Charles. The dissociative technique used, whether Ouija board or table-tapping, allowed the words to flow. Materer (2000) noted that Merrill was a reluctant prophet in the true tradition, whose angelic voices pressed him on until the tale was told. Harvard professor Daniel M. Wegner (2003) said it this way: 'The belief in outside agents who influence a person's actions can so muddle the per-ception of conscious will as to promote bizarre dissociations of perceived authorship in the form of trance channeling, spirit possession, and dissociative identity disorder' (Wegner 2003: 68). No matter how the poet tells his tale, its unwilled nature appears to be uncanny.

Lurie suggested a psychological force behind the joint Ouija board sessions. Jackson's hand in them, using his 'subconscious mind', gave back purpose he had lost when he failed to publish on his own (Lurie 2001: 107). It also kept the two men together, as the spirits favored their union (while also suggesting they take other lovers). I agree that Jackson's role was essential to the making of *Sandover*, even though he was often uneasy, fearful and drained by a process he consciously resisted and once described as 'schizophrenic' and 'very, very egocentric' (letter to J.D. McClatchy, Qtd. in Lurie 2001: 117–18).

JM's 'ex-shrink', Dr Detre, termed the collaboration a *folie à deux*. Rather than a double madness, it was a force that brought them together, if not perfectly, like Philemon and Baucis, the legendary couple who recognized the gods and were rewarded with metamorphosis into interlaced trees: 'David and I lived on, limbs thickening / For better and worse in one another's shade' (*CLS*: 41).[10] But there were also dangers. One page later in the long poem, JM will re-experience his traumatic 'death' in a former life, as Catherine did repeatedly in Dr. Weiss's office.

Most tellingly, the 'spirits' informed JM and DJ that the Red room was where the 'POWER LIVES… THAT WILL PERMEATE YOUR MIND'. Directed to close their eyes, both men experienced a curious physical phenomenon indicating its neural provenance. First, JM:

> We do. A faint, pulsing tremor begins
> In my left arm, shoulder to fingertips
> Poised on the cup I meanwhile judge to be
> Moving slowly, slowly, from 1 to 0
> (Passage that takes a minute, more or less)
> Three times. Then suddenly a sense of — yes —
> Whiteness on my left side. Whiteness felt
> Against my cheek, along my forearm, like
> A wash of alcohol that as it dries
> Refreshes. The cup rests. Open our eyes?

Meanwhile, DJ's 'left hand all this while unawares / Pressed flat against the Board — how did that happen?' (*CLS*: 213). As DJ himself described it: 'I suddenly felt this terrific pain in my left hand. It was really extraordinary. In fact, I couldn't move, was just literally pinned to the board, Jimmy with right hand writing. And this cup just moving like a sadistic maniac'. It was demanding 'three more years of our time, and that we… stop fooling around' (McClatchy 1979: 34). Rather than the workings of their unconscious minds, I would say their conjoined right hemispheres were working autonomously against their left.

A panoply of poetic predecessors[11] appeared in the poem — Homer, Dante, Hugo, Rimbaud, Mallarmé, Eliot, Yeats and W.H. Auden. Wallace Stevens and Jung were included for their views on the destructive Anima and God and the unconscious as one. While not a poet, Proust, the subject

---

[10] Hugo also had a ritual of pointing out two trees with interlacing branches on his daily walk with his mistress, Juliette Drouet, referring to Philemon and Baucis. André Breton would find this '*cérémonie*' incredibly touching in the poet predecessor he so admired, as cited in *Nadja* (Breton 1928/1964: 12–13). As Bloom said, predecessor poets are gods to their successors.

[11] As Lurie noted, there were no novelists, only poets like JM. Nature calls up the self-similar, which instructs metaphorically.

of Merrill's undergraduate thesis at Amherst, would play a role. Spirit talk would be in small CAPITALS, while JM's and DJ's case was lowered. Proust, for example, was 'ABOVE ME   A GREAT PROPHET THRONED ON HIGH' (*CLS* 76).[12] In right-hemispheric fashion, identities destabilized, multiplied and divided according to the needs of the evolving word. As the theory expanded, 'spirit' entities morphed, climbing their way up the spiritual hierarchy.

The tone of Hugo's voices was humorless, arrogant and pompous, unlike Merrill's, who were more often witty and clever. Vendler also noted how they differed from Dante's and Yeats's 'gloomy reverence for their guides' (Polito 1994: 139). Merrill, who did his own doctoring, criticized both Hugo and Yeats for altering *their* spirits' messages.[13] In *Mirabell: Book 2*, Hugo is called an 'OVERSTIMULATED SCRIBE', qualified as 'A GREAT PROBLEM OF FRENCH CULTURE   OVERHEATED   FAME ALWAYS ON THE BOIL' (*CLS*: 144). Merrill claimed he was 'half trying / to make sense of *A Vision* / When our friend dropped his bombshell: "POOR OLD YEATS / STILL SIMPLIFYING"' (*CLS*: 14). Yeats both inspired *The Changing Light at Sandover* and credited superior spiritual wisdom to his successor poet. Downplaying the predecessor's efforts rings a Bloomian bell, which Bauer (2003) spelled out in his book.[14] After all, if spirits were correctly representing the afterlife, they all would say the same thing, would they not? That is, unless we create our own afterlives based on our past experiences, prejudices and preferences, which seems to be what Merrill did with his reincarnation scheme.

According to Stephen E. Braude (2003), who has researched reincarnation thoroughly, claims are particularly abundant in cultures that fully support the belief system, which is what Jaynes would have predicted.[15] Merrill adhered more to his predecessor Yeats's model than to any foreign culture's; nonetheless, the poem exemplified Braude's conclusion that 'spirits' assumed new bodies most often after suffering a violent death by

---

12   Recall that the 'above me' spatial metaphor had been used in Hugo's séances as well.

13   The ninth volume of the Hugo's *Oeuvres Complètes*, which included the existing transcripts of the talking tables, had come out in 1968; Merrill was obviously knowledgeable about their contents, referring to the fact that 'Characters from fiction and full-fledged / Abstractions came to Victor Hugo's tables' (*CLS*: 143).

14   Bauer (2003) also considered Yeats 'a vital presence' in Merrill's work, both 'appropriated' and 'resist[ed]'. Merrill was a 'poet for whom reading is a ghostly dialog between distinct yet "co-operating" lives' (Bauer 2003: 108–9). Yeats remained deflated and mostly 'wordless' in the poem because he had been covertly but 'fully used' in *Sandover* (Bauer 2003: 139).

15   Braude (2003) names 'south and southeast Asia, western Asia, west Africa, Brazil, and Alaska' in particular' (Braude 2003: 177).

fire or drowning.[16] Braude, who accepts the possibility of survival after death, at least for a time,[17] states further that loss or separation in this life often plays a role in those who experience co-consciousness with persons from previous lives, whether real or dissociative, because of a psychological need. Are those who contact spirits 'gifted' or 'afflicted' (Braude 2003: 225)? I would say both, with the further possibility, as Conforti suggested in Chapter 4, that a field of complexes might persist through time.[18] '[I]f you wonderd why we come & more, why one of the white shd risk himself: yr field is yes a kind of anchor point of heaven. O scribe, o hand u have paid yr dues again & again   for who living welcomes the dead?' (*CLS*: 257–8). The notion of 'great magnetism' has been proposed in many 'spirit'-produced texts (*CLS*: 265).[19]

Merrill's consternation about his place in the visionary line came down to a fundamental question: 'What part, I'd like to ask Them, does sex play / In this whole set-up? Why did they choose *us*? / Are we more usable than Yeats or Hugo, / Dotters on women, who then went ahead / To doctor everything their voices said?' (*CLS*: 154). In fact, Hugo's, Yeats's and Merrill's otherworldly contact did have a common impetus: *childlessness*. The Hugo family contacted the dead to find their daughter, Léopoldine, lost with her new husband in a tragic boating accident. The poetry, the abstractions, the illuminations from illustrious predecessors followed on from that initial search, once what was lost was found in heaven. As we saw, Yeats married the much younger Georgie because at nearly 49 he had a book, but no child. She, in turn, seemed to orchestrate the whole endeavor to encourage their sex life in pursuit of a messianic offspring – in the reincarnation of a child lost to a distant relative in the past.

---

[16] Many of the better claims come from India, per Braude. It is not surprising that Merrill's dear friend, Maria Mitsotáki, would be 'reborn' there and revered even in infancy.

[17] Summing up, Braude says: 'And I think I can say, with little assurance but with some justification, that the evidence provides a reasonable basis for believing in personal postmortem survival. It doesn't clearly support the belief that everyone survives death; it more clearly supports the belief that some do. And it doesn't support the belief that we survive eternally; at best it justifies the belief that some individuals survive for a limited time' (Braude 2003: 306). Jung came to the same conclusion after his own NDE. Merrill referred to Jung's NDE in *CLS*.

[18] Braude cites C.D. Broad: 'If something survives death, it might be like... a persistent vortex in the ether, carrying modulations imposed on it by experiences had by the person with whose physical body it was formerly associated as a kind of "field"' (Broad 1962: 430; cf. 419).

[19] When I asked the 'angel' why she had come to my friend, her reply was similar: 'she is like a magnet'.

When Merrill's psychiatrist asked him why he thought this 'psycho-roulette' was going on, the poet offered: 'Somewhere a Father Figure shakes his rod / At sons who have not sired a child?' (*CLS*: 30). The 'genetic angel' had struck when he was 'nearly thirty and not yet a father', living in Stonington with David, far afield from the New York social scene. By the following summer, Merrill would renounce the 'Annunciation' and fill the house with 'Ephraim and Company, who were prepared, like children, to take up as much of our time as we cared to give, but whose conversation outsparkled Ravenna, and who never had to be washed and fed or driven to their school basketball games' (Merrill 1994: 203). His 560-page poem, created from his Company's chatter, would be a paean to the superiority of creative homosexuals, not the patter of little feet — a literary defiance to 'nature's law: mate propagate & die' (*CLS*: 229).

The child connection comes up again as attempts are made to 'place particular souls [Ephraim's "representatives"] with women they choose'. JM offers up his niece, Betsy, when a 'strong sane woman' was needed to give birth to one of Ephraim's charges, drowned in the bathtub in his previous life. DJ offers up the wife 'on the nest' of an ex-roommate (*CLS*: 19–20). If souls can be implanted at the will of the spirits, the parents' gender no longer matters. Ephraim uses both nesting moms, but not before the sixth month, to assure viability. In JM's and DJ's mind, they have orchestrated two births, collaborating with their otherworldly matchmaker. Unfortunately, in DJ's case, he had supplied the wrong name.[20]

Along with a need to justify their lack of propagation, a series of significant deaths fueled the Ouija board quest: Hans Lodeizen, JM's Dutch poet friend, dead in 1950; JM's father in 1956; Maya Deren, filmmaker and once-possessed hippie, in 1961; poet and father figure W.H. Auden in 1973; Maria Mitsotáki, both surrogate mother and muse, in 1974; and DJ's own mother. Others among the *Dramatis Personae* included: John Clay, a clergy-man, dead in 1774, now 'patron' to DJ; Kinton Ford, editor of Alexander

---

20   Braude also describes an interesting case where a woman with maternal yearnings considers adopting a child and marrying an older doctor with whom she claimed to have had a relationship in a former life. Her sensitivity to the fear of never having a child was so strong that when she read T.S. Eliot's 'The Wasteland', she felt barren herself and experienced 'abortional bleeding'. The next day, after teaching a poem about a boy on a burning deck, she reportedly thought she should have such a boy and then bled again. Her dissociative problems revolved around her desire to have a child and her lack of a suitable mate (see Braude 2003:103–6). This anecdote shows how powerful the child connection can be when linked to resonant language. In Merrill's case, it worked the other way: the guilt about not having a child for his parents translated into an epic poem disculpating him.

Pope's works, in 1843, now patron/couplet producer to JM; Rufus Farmetton, a previous incarnation of JM, whose death he re-experienced, in 1925. The last name given was W(illiam) B(utler) Yeats (1939), the poetic predecessor of greatest import, whose 'maze of inner logic, dogma, dates' JM and DJ were 'half heartedly' exploring before Ephraim appeared in their lives. The partners declined to edit WBY on behalf of 'someone up there' who erroneously thought they might be able to do so (*CLS*: 11–14).

Merrill was completely forthright about his Ouija board odyssey and gave many illuminating interviews to explain the process first-hand. To Vendler, he differentiated his long poem from his previous ones for the 'unprecedented way in which the material came. Not through flashes of insight, wordplay, trains of thought. More like a friend, or stranger, might say over a telephone. DJ and I never knew until it had been spelled out letter by letter' (Merrill 1979: 12). He granted that the material, while not very real, was indeed imaginatively real, with 'proportions broader and grander' than his own imagination (Merrill 1979: 12). He admitted that he could not have written the work without the aid of the Ouija board, which allowed each transmission a spaced-out transit time that a 'saint or a lunatic' would have received 'in one blinding ZAP' (Merrill 1979: 13). In the poem itself, he said its lettered delivery was in no way 'some holy flash past words' (*CLS*: 63). 'If the spirits aren't external, how astonishing the mediums become!', he told Vendler. Finally, he quoted Hugo saying 'of *his* voices that they were like his own mental powers multiplied by five' (Merrill 1979: 13).[21]

Merrill's Psychological Need

In his memoir, *A Different Person*, Merrill wrote that between the ages of 9 and 10, when his mother had little time for him, he transferred his love to his governess, 'Mademoiselle'. At 11, when his parents were divorcing and he needed Mademoiselle most, his father decided his son needed 'masculine supervision' (Merrill 1994: 130–1) and hired a young Irishman to replace her. Longing for the beloved Mademoiselle, young Jimmy rebelled and the Irishman quit after two months.

Looking back on his boarding school days, Merrill noted that most of his friends' fathers had either drifted off or committed suicide and 'the trauma seemed in every case to have quickened the child's imagination' (Merrill 1994: 9). The 'quickening' may rather have resulted from a depressive state brought on by loss, which fostered right-hemispheric poetry. When Merrill asked Ephraim why '[m]ust *everything* be witty', the

---

[21] Hugo was actually referring to his own son, Charles, who otherwise could not have produced such wisdom (see Hugo 1864/2003).

'spirit's' response is telling: 'AH, MY DEARS / I AM NOT LAUGHING   I WILL SIMPLY NOT SHED TEARS'.

Merrill's poetry writing began in prep school and continued through college. At Lawrenceville, he had studied all of English poetry and also read Baudelaire and Verlaine. At Amherst, Merrill met Robert Frost, who kindly critiqued the budding poet's work, fully acknowledging his indebtedness to Rilke and Yeats. But his best friend, Freddy, also strongly influenced him 'in the vast chamber full of [predecessor] voices' (Merrill 1994: 15). Amherst English professor, Kimon Friar, mentored Merrill and became his lover. Merrill made clear that *Sandover* fulfilled *Kimon's* dream of writing a great work 'based on Yeats's system: spiritualism, the phases of the moon, the gyres of history'. While once considering it 'megalomania' to attempt a work on par with Dante, Milton, Rilke and Pound, Merrill deemed his project '[l]onger than Dante, dottier than Pound, and full of spirits more talkative than Yeats himself might have wished'. He inscribed Kimon's copy of the heavy tome with this succinct message: 'Dear Kimon, who'd have thought? You would!' (Merrill 1993/1994: 27).

Merrill considered his father a benign figure. In fact, he would compare his father's inconstancy in love to his own. At the time of his memoir Merrill wrote: 'To this day he remains an almost perversely mild and undemanding presence in my thoughts' (Merrill 1993/1994: 42). In *Sandover*, when learning his father had died, JM expressed '16FOLD LACK OF EMOTION', as a centuries-old Zen 'priest' approvingly informed Ephraim. When the father's 'spirit' did get through to JM, he was 'incredulous' the board worked for his son, having failed when he had tried it after his own mother's death. Pronouncing his love for all his wives and children, the paternal spirit scurried off, looking forward to his next life. The irony of the head of Merrill Lynch reborn to a greengrocer in Kew most likely pleased his son. Although the elder Merrill's new name and address were spelled out, JM was forbidden to interfere. Most endearingly, Merrill's father had once gathered his son's early poetry together into 'Jim's Book'. When Merrill's actual first poetry volume came out, his father ordered a hundred of the 990-print run for his friends and business partners. 'One tenth of the tiny edition doomed to oblivion, at a single stroke', Merrill wryly commented (Merrill 1993/1994: 63). Two years later it was not sold out.

Merrill's ongoing emotional battle was with his mother, who disapproved of his sexual preference. After his affair with Kimon Friar, Merrill followed his mother's 'earnest wish' and saw a therapist. His father, seemingly less savvy about his son's proclivity, had only once remarked to him on the occasion of his induction into the army: 'Never let another man put his hand on you' (Merrill 1993/1994: 90). Later, Merrill learned his father had requested a joint interview with his Rome-based doctors to talk about his homosexuality. Whether parental disapproval or outright admonitions, Merrill considered his treatment mild and not particularly meaningful since 'hundreds of thousands of parents — not just mine — must have spent the forties and fifties urging secrecy and repression upon their queer sons' (Merrill 1993/1994: 151). Jackson's father, who had been critical of his son in life, was more accepting after death: 'YOU 2 ARE OK' (*CLS*: 103).[22]

Throughout his memoir, Merrill seemed to question the origins of more pervasive problems — lack of a secure sense of self, his inability to be alone, feelings of unworthiness — attributing most to his early family situation. His sexual orientation played a prominent role in his memoir, especially his mother's efforts to 'make him a different person' or at least hide his lifestyle and troublesome emotions from public view (Merrill 1993/1994: 97). On one occasion, while he was in Rome, his mother destroyed, at his direction, not only a box full of letters to him stored in her home, but also others, from friends and lovers, in his own apartment. The 'BURN THE BOX' motif occurs repeatedly in *Sandover*, most likely referring to a box of Ouija-board transcripts he and Jackson had nearly burned as well. Either way, Merrill emphasized his mother's overbearance, saying her act 'left me with little evidence of having been loved by anyone, except her' (Merrill 1993/ 1994: 97). Despite her apparent absence on *Sandover*, his mother loomed large over the trilogy, having engrained in him a love of art, mannered expression and camouflaged emotion: '…Because of course she's here / Throughout, the breath drawn after every line, / Essential to its making as to mine' (*CLS*: 84).

While understanding the link between trauma and creativity, Merrill nonetheless relativized that connection, claiming 'something of the kind awaits every child on earth'. Without the parents' 'imprint of (imperfect) love, the self is featureless, a snarl of instincts, a puff of stellar dust' (Merrill 1993/1994: 155). Nonetheless, his sexual preference and 'the Divorce' came up again and again in his work, leading him to ponder why his father appeared in all the men he loved and if his 'life has been less a flight from the Broken Home than a cunning scale model of it' (Merrill 1993/1994: 154-5). Studying an early photo of himself showing 'a sissy at six, posed, hands folded and ankles crossed, at the slide's foot', Merrill associated the image with a Michelangelo statue, unfinished and abandoned because of an internal flaw. Disculpating his parents, Merrill thought perhaps his marble was already split before their divorce; and, if it had not been so, 'this

---

[22] Fathers recanting in the afterlife are typical of spirit transmissions, according to Sword (2002). Mediumship is not all about slaying 'the Author and the Father', but rather a response to the loss of loved ones and one's own eventual demise, palliated by 'cheating death through language' (Sword 2002: 48).

inward, famished understudy for creative Love would never have come to light' (Merrill 1993/1994: 139).

Despite his disavowal of their role, a possible genetic predisposition to increased right lateralization, the trauma of the broken home and his parents' opposition to his homosexuality may well have destabilized Merrill's sense of self, rendering him a kind of Keatsian 'chameleon'. His 'spirit' contacts amply and whole-heartedly supplied the bolstering he needed. Their overblown and capitalized praise countered his own reductive sentiments, such as, 'Already I take up / Less emotional space than a snowdrop' (*CLS*: 89). Merrill had long ago concurred with Freddy's remark that traumas are a constellation and a blessing that give life meaning.[23] Both prophetic and practical, Freddy would give JM his first Ouija board, a prelude to inner revelations and creativity beyond his wildest dreams.

Mirabell's Books of Number

In Part II of *The Changing Light at Sandover*, a change in home décor led to changes in the spirits' chatter. Merrill and Jackson enlisted their friend, Hubbell, to create wallpaper that would incorporate fans, clouds and bats from the design in their new carpet, while they vacationed in Greece. Like-wise, a beaded curtain with a swaying bird, picked up in Istanbul, now hung between the red and blue rooms, and may well have conjured the peacock Mirabell's appearance (see Merrill 1994: 244–5).

While JM and DJ thought they had closed the book on Ephraim and Friends not long after the incident with DJ's pinned-down hand, more deaths—DJ's mother, Mary; Maria Mitsotáki and Auden's friend and collaborator, Chester Kallman—brought the partners back to the Board, unveiling new revelations. Children from past lives meet up with Mrs. Jackson. They learn that the dead only see themselves in the mirror of a living mind and spirits choose the living to be 'vehicles' for scientific and artistic breakthroughs. Told he must write 'POEMS OF SCIENCE', JM opened a 'biophysichemical' textbook, despite 'pity and dread', in preparation for his inspiration. Quantum physics met the muse Urania, in whose face JM, like Keats before him, *almost* knew 'What's matter', despite a difficult learning curve (*CLS*: 110–11).[24]

---

[23] In his memoir, *The Blessing*, Gregory Orr made clear that his accidental killing of his younger brother, his problematic parenting and his self-sacrificial lifestyle all led to his eventual birth as a poet.

[24] Yenser (1987) said that Merrill had been reading *The Lives of a Cell* by Lewis Thomas, who was both a medical doctor and a poet. Thomas wrote that man was separating himself from Nature and becoming a 'lethal force' for the planet; man was not singular, but rather inhabited by many tiny organelles that orchestrated inner processes. The Earth too was a single cell

As with most dissociative texts issuing from the right side, the language was both poetic and religious, despite the fact that religion had not been the friends' strong suit beforehand. Early on, when Ephraim asked the partners if they were 'XTIANS', they weakly 'guess[ed] so' (*CLS*: 8). Following the Hugolian and Yeatsian mold, a reincarnation scheme evolved as a highly selective process. Most patrons crammed their representatives' souls with knowledge between incarnations — the only time they could intervene. However, Ephraim, having discovered JM through his first connection with the fire victim Simpson, could express his knowledge during this life via the Ouija board.

What Merrill learned did not resemble in the least the lessons of his predecessor poets. Life is not a school, in the Keatsian scheme, but rather a 'disastrously long' vacation from the patrons' ministrations (*CLS*: 10). This system did not reward goodness with weightlessness and punish evil with entrapment in dense matter, as with Hugo. Rather, 'density' was a precious commodity, like a pregnancy of intelligence and creativity, fired by complex equations in the soul-cloning process of a celestial Research Lab (R/Lab). The majority of the world's population had been reincarnated like 'THE SAME OLD HORSE RESHOD   WE THERE4 LEAVE TO THE BLACKSMITHS THEIR 3.5 BILLION LIVES'. Whereas, 'IN THE HUSHD SANCTUM OF THE R/LAB' the 'HAPPY FEW' would be 'IMPROVE[D] HORSE & SHOE ALIKE' (*CLS*: 145).

God is not One: '2 GODS GOVERN   BIOLOGY & CHAOS', pitiless, natural forces that *require* plague, war, suicide and natural disasters to 'SLOW THE CROWDING' and mine raw materials to clone superior souls. With five million needed each day, the bats had recourse to animal souls, who brought 'JOIE DE VIVRE' and plant souls who contributed 'GREAT VITALITY' (*CLS*: 151). Unapologetically, 'THE HITLERS THE PERONS & FRANCOS   THE STALINS' play a role in the relentless pruning and recycling project (*CLS*: 188). Wystan, Maria, JM, Mirabell and DJ merge to become a holy force of five: water, earth, air, fire and the shaping hand of nature. All five would do the 'V' (or *vie* = life) 'WORK GUIDED BY HIGHER COLLABORATION' (*CLS*: 162). Intermingling and authorial meddling become the order of the day. By page 217, we learn that Yeats was moving DJ's hand all along and Rimbaud ghostwrote 'The Waste Land' — the resisted foster child of T.S. Eliot. Resistance itself becomes a cosmic theme by the end of the epic poem.

The impatient forces that had hijacked the Board from Jackson were fallen angels — Cain's sons, forces of 'Chaos' — who intervened to explain how the Fall had created the negative energy of black holes and their role

---

in which everything was interdependent. Merrill's bats would say: 'the whole greenhouse / is but a cell, complex yet manageable all matter / there4 is part, of that cell' and 'god b[iology] is not / only history but earth itself he is the greenhouse' (*CLS*: 210).

in tending the fires of mind that feed God Biology. 'MYND AND NATURE WEDDED' were JM's mind and words and DJ's feeling nature. JM suspected that the negative flames that lapped around them now were 'largely metaphorical'. These Mephistopheles concurred, 'INDEED JM WE HAVE ALWAYS SPOKEN THROUGH THE POETS' (*CLS*: 113–14). Yet, their highly qualified dead friends relating through the Board were merely fodder, cloned not bred, to populate the Earth with their purer stuff.

Where Hugo's focus was a European Union and justice for the poor and women, Merrill's scheme progressed toward a Utopian intelligentsia, a 'PRECIOUS NUCLEUS OF MINDS' (*CLS*: 119). Language is pronounced the life raft that saves civilization. JM is crowned a scribe in the 'UNBROKEN CHAIN HOMER DANTE PROUST EACH WITH HIS SENSE OF THE MINDS POWER ITS GENERATIVE USES' (*CLS*: 119–21). With Mind over Matter, overpopulation becomes the principle concern. The bats, speaking from the 'UNFERTILE WOMB OF CAVES' castigate Pope John Paul for his position on birth control.

Clearly, the give and take between the partners and their interlocutors showed that anything they had seen, read or thought could make its way into the creative storytelling emerging from the Board. For instance, after JM and DJ visited the stones of Avebury, followed by the treasure of Sutton Hoo at the British Museum, the bat creatures claimed they had seen them there. They talked of anchor sites attached to glowing stones and explained that buried Anglo-Saxon boats had been used as offerings. The creative mind recombines any associative materials at hand. Spirit mediums typically can create entities based on things or people they have seen or read about or, possibly, culled from their séance sitters' minds. Pearl Curran, a high school dropout with little interest in literature, was born in Illinois in 1883. Her trance personality, Patience Worth, a seventeenth-century English woman, was able to produce dissociative texts, including poetry, novels, short stories and plays, some of which were written in an archaic dialect.

Given Curran's feat, it is not surprising that Merrill, with his intelligence, education, cultural repertoire and voracious reading in mythology, poetry and psychology could produce his Ouija board cosmos in collaboration with his literate and cultured partner. In some cases, visual imagery, alliteration, assonance and metaphor (M) spurred their dissociative imagination.[25] As Merrill declared: 'Indeed & not accidentally M is at once our method & the midpoint of our alphabet the summit of our

---

[25] The right hemisphere processes vowels and the left consonants (Carreiras and Price 2008), which explains assonance. Stressing a word's beginning consonant for alliteration is also right hemispheric (Kane 2004). JM's spirits did not *speak* in rhymes (a left-hemispheric acoustical matching); rather, he consciously created them along with the meter.

rainbow roof in timbre the mild meridian blue of muse & musing & music the high hum of mind' (*CLS*: 222). The countervailing accent on science might have been another left-hemispheric contribution to the poem's puzzle. Merrill was indeed 'reaching after fact & reason' to prove his case.

At the end of Part II, heaven is explained as both 'reality & a figment of imagination... a new machine which makes the dead available to life' (*CLS*: 260). It exists as a recycling plant for the 'newly dead' who 'howl like dogs in a pack on their way from carnage' until 'they are calmd into usefulness' (*CLS*: 264). Hell is real and as boundless as the negative imagination. The voices JM and DJ had been hearing, even Ephraim's, were not singular, but 'composite', as formulaic as new souls composed for the good of mankind in the Research Lab. In the final revelation, the sun strikes the first cell into life with its energizing rays and identifies 'god' as 'the accumulated intelligence in cells since the death of the first distant cell'. The Archangel Michael, 'guardian of the light', announced more meetings to come and imperiously commanded them to 'look! look into the red eye of your god!' (*CLS*: 276). Thomas (1974) had called the sun the 'unremitting, constant surge of energy' that fueled the planet (Thomas 1974: 25).

Scripts for the Pageant

The last part of *Sandover*, which is divided into three parts, identified as the 'Yes', '&' and 'No' symbols on the Ouija board. A new hierarchy to replace traditional religion emerges: God Biology and his twin, Nature, also known as Psyche and Chaos. The elemental archangels, Michael (Light), Emmanuel (Water), Raphael (Earth) and Gabriel (Fire), along with the nine Muses, a senior officer in Gabriel's legions, and Mirabell, all take part. 'The Five' senses are also represented: Akhnaton (sight), Homer (hearing), Montezuma (touch), Nefertiti (taste) and Plato (smell). Four religious leaders are included, the Buddha, Jesus, Mercury (who does not speak) and Mohammed, along with the beloved and honored dead ones: W.H. Auden, Maria Mitsotáki,[26] Maya Deren, Hans Lodeizen, assorted writers and musicians, Ephraim, and a unicorn named Unice.

In 'YES', Wystan describes the archangel Michael as 'A CUMULUS MODELED BY SUN TO HUMAN LIKENESS... FACE OF THE IDEAL PARENT CONFESSOR LOVER READER FRIEND & MORE' (*CLS*: 286). The facial likeness recalls the early

---

[26] Merrill (1994) said he was not surprised that Wystan and Maria, aka Maman, became the 'leading lights of Sandover'. He described her as: 'Dressed in her eternal black, wreathed in the smoke of her eternal Gitane, Maria is the closest I'll ever get to having a Muse. There is no one saner or more sympathetic, more in love with overtones, quicker to register anything said or left unsaid' (Merrill 1994: 230–1).

developmental right-hemispheric connection to the loving parent's face
and subsequent important dyads. JM's and DJ's spirits express the same
sort of grandiosity as Hugo's had proclaimed: 'INTELLIGENCE, THAT IS THE
SOURCE OF LIGHT. FEAR NOTHING WHEN YOU STAND IN IT   I RAISE YOU UP
AMONG US   HAIL   HAIL' (*CLS*: 296).

Elijah decries overpopulation: 'FOR IT IS GOD'S WILL THAT HIS CHILDREN
REDUCE THEIR NUMBERS' (*CLS*: 307). WHA, senior poet and surrogate father,
and mother Maria Mitsotáki tearfully defend against a chaotic end to Earth.
Consigning each sense to a highly significant figure, whether real
(Akhnaton) or mythic (Mercury), Merrill creates his own mythic/meta-
phoric story to *make sense* of the complex reality of the origins of the
universe and the fate of life on Earth. The right hemisphere, I might add,
coordinates sensory input from the environment and is active in creating
mythic stories (D'Aquili and Newberg 1999).[27]

Two more deaths occur. First, George Cotzias (called 'GK') sits out his
own funeral with his Earth-bound friends and witnesses the firing of lab
souls in the Other, a process done 'WITH A GENIUS FAR / BEYOND THE DULL
TRANSMISSION OF A GENE BY EGG & SEMEN' (*CLS*: 375). Next, Robert Morse
(RM) dies in his sleep. With these deaths comes the knowledge that loss is
necessary for the creativity of the living: 'The buffeting of losses which we
see / At once, no matter how reluctantly, / As gains. Gains to the work. Ill-
gotten gains' (*CLS*: 376). Affirmation of reincarnation dulls the sting of
Robert's loss: 'B4 THE FULL MOON SWEEPS US OFF TO 5 / RM WILL COME ALIVE!'
(*CLS*: 376-7). The theme of profit from loss is repeated in the epilog, '*Coda:
The Higher Keys*', as Morse describes life from within a new womb. Now
named 'Tom', and lame like Byron, Morse is destined to become a com-
poser: his wound becomes his blessing. Spiraling into manic meaning,
nothing in the dictation was as it seemed and Yeats was the 'WORDLESS
PRESENCE' behind DJ (*CLS*: 424).

In 'NO', the Scripts become a stage on which to plead for Earth's
salvation. Gautama claims he preached 'GREAT EARTHBOUND CEREBRALITY',
but the Hindus set in motion 'A PINWHEEL OF SPUTTERING LIVES / EACH MORE
USELESS THAN THE LAST...' (*CLS*: 443). Jesus decries the '...DEAD SOUND, MY
NAME, IN HALF THE WORLD'S PULPITS' produces, and begs to come to Earth
again to 'IMPLORE / WRETCHED MAN TO MEND, REPAIR WHILE HE CAN AMEN'

---

[27] '[T]he nondominant hemisphere... generates the perception of sensory
input as a whole rather than as a string of associated discrete elements. We
propose that the cognitive unification of logically irreconcilable opposites
presented in the myth structure (such as god and human in a solar hero or a
Christ figure) represents a shift of predominating influence from the left
hemisphere to a predominant influence of the right hemisphere, which
allows the antinomies to be perceived in a more unitary or integrated mode'
(D'Aquili and Newberg 1999: 87).

(*CLS*: 444). Gabriel tells JM he will be given 'NEW MATERIALS… FOR A NEW FAITH, ITS ARCHITECTURE, THAT FLAT WHITE PRINTED PAGE TO WHICH WILL COME WISER WORSHIPPERS IN TIME' (*CLS*: 446). This message meets up with earlier ones saying 'SCIENCE, POETRY & MUSIC' are now the three mono-theisms' (*CLS*: 239) and 'HEAVEN… IS THE SURROUND OF THE LIVING' (*CLS*: 59).

Mohammed, waiting shyly by, is '…very much alive… but arab faith & politics / combine into a fairly heady mix' (*CLS*: 446). At the next angelic lesson, this 'simple man' shows his face, confused that such a vision would come to him, 'who could not read [but] spoke' and 'it was easy'. Mohammed was following orders: 'thin out your race and keep it thin with bloodshed, for you sit on time made black' (*CLS*: 448–50). Overpopulation, along with Earth's resources plundered by too many oil wells, is breaking 'the whole frail eggshell' (*CLS*: 453), not to mention the danger of a nuclear explosion 'underground siberia' (*CLS*: 462). The nuclear power plant Chernobyl disaster occurred in Ukraine in 1986.

The Final Hurrah

When JM began *Ephraim*, Maria Mitsotáki (MM) had just died. We now learn she was actually Plato, vacationing from the density of men, joyously living his last life as a woman. All nine Muses rolled into one, Maria is Mother Nature's child and the one who brought JM to Archangel Michael for the benefit of all humanity in the first place. Yeats appears one last time with new versions of his cones and gyres. He declares DJ 'NEARLY AS GOOD AS A WIFE'. Called a 'CROUCHING ELDER SCRIBE', Yeats climbs out from under DJ's palm, 'whips out pince-nez' and addresses a brief 'SPEECH… after LONG SILENCE', then crawls back under again (*CLS*: 486).

Amid fanfare and MARCHING TUNES, Mother Nature declares '…A LASTING RESOUNDING YES / TO MAN, MAN IN HIS BLESSEDNESS' — but this blessedness is diminished by overpopulation (*CLS*: 489). Maria, now reborn as a miracle-working Indian baby, keeps the 'WRETCHED POPULATION / HAPPILY AT STARVATION LEVEL' (*CLS*: 542). Mother Nature was Ephraim too, whom she possessed in his Tiberius days. 'OUR MOTHER'S WAYS TEND TOWARD ECONOMY / It was You, You always, the whole time' (*CLS*: 552). 'NOT THE MOMENT QUITE / TO GOSSIP BUT THERE'S ONE THING YOU SHOULD KNOW. / THESE WORKS, YOU UNDERSTAND? THAT OTHERS 'WRITE' / … / ARE YET ONE'S OWN…' (*CLS*: 557–8). All is One. Self is Other, Other Self.

As the final curtain falls, the doorbell rings. Vasíli, a Greek novelist friend, brings news of Mimí, his *childless* wife who has died in Rome. Another death and an empty womb spur a convivial reading to the assembled 'spirit' guests. The poem begins again: 'Admittedly…' (*CLS*: 560). The triumphant poet had seeded a message to save the Earth while justifying his sexual orientation and his childlessness. Maria, the surrogate

Mother, had confirmed Merrill's place in the Pantheon of Great Poets. The Predecessor had crawled beneath Jackson's hand. Opening the box of his own mind, Merrill's Ouija board odyssey of return had tossed up costumes leading to his own true Self. 'Sense, comfort and wit' had indeed been served up (*CLS*: 100).

Chapter Eight

Right-Hemisphere Imbalance versus Espousal: Sylvia Plath and Ted Hughes

My dearest Sylvia – two letters from you today. The first
melancholy sad, the second its antithesis.
  – Ted Hughes

I wonder about the poems I am doing. They seem moving,
interesting, but I wonder how deep they are. The absence of a tightly
reasoned and rhythmed logic bothers me. Yet frees me.
  – Sylvia Plath

Their marriage floundered and their art did not.
  – Diane Middlebrook

Signs of Atypical Lateralization

Certain outer signs in Sylvia Plath's appearance, as well as in her creative life, point to atypical lateralization. While women in general have been shown to have greater bilateral representation for language than men (Vikingstad *et al.* 2000), Plath's bilaterality seemed to go beyond the norm. Even as a child, she regularly used words and visual imagery together (Connors and Bayley 2007). As confirmed in many available photos, we see how, with arms folded across her chest, both hands go up; with both arms extended, neither was preferred. Despite being right-handed, she always parted her hair on the right like a left-hander, as did her mother.

Plath attempted sounding out speech at 6 to 8 weeks old, according to her mother, and acquired the prosody of foreign languages easily (French and, to a lesser extent, German). Prichard *et al.* (2013) say the advantages of bilateral dominance may include better and earlier episodic memory retrieval, superior face memory, better word retrieval in foreign languages and more creativity via divergent thinking. Fiction and poetry became equally important to Plath. Both were crafted from her own experiences and theoretical readings, as well as from the influences of admired writers and poets. Jarring metaphors were not confined to her poetry; her prose was highly metaphoric as well. As an adult, she was an artist whose fine-detailed, linearly oriented drawings accented concrete objects and places in her physical environment, whether at home or traveling abroad.

Along with the advantages above, we must keep in mind that mixed dominance is a predisposing factor for both mental illness and paranormal beliefs. Plath's written creations drew heavily on her extreme emotional states, her traumatic experiences and unresolved losses, especially, and ultimately, in times of great stress. Her artistic creations, on the other hand, reflected calm and order for the most part, suggesting the role of the hypothesized less dominant, yet more positive, left hemisphere. She claimed, and Hughes confirmed, that drawing calmed her down. If we compare Plath's art to Hugo's, we see how hers depicted external reality in upright broad strokes. Hugo's art relied heavily on dream-like pen and ink washes, often with his name or initials intertwined within the picture. In addition, he *chose* to draw with his left hand, despite being right-handed, to access unconscious inspiration. Hugo *showed* his turbulent emotions manifested in the outer world; Plath *controlled* her emotions, drawing a calm, ordered external world separated from her sense of self.

Ted Hughes appears consistently right-handed in his photographs: his right hand is always posed under his chin in headshots. He holds his pen, wine class and fishing rod with the right; hands folded, his right thumb is on top; legs crossed, the right is on top; his hair is always parted on the left for right-handed grooming. Yet, in photos where he folds his arms across his chest, the left hand goes up, showing a measure of mixed dominance.

Mainly, his reading preferences[1] — his non-fiction writings centering on the minds and works of admired poets — his unswerving devotion to own his poetic vocation, along with his symbolic dream-life and paranormal beliefs and practices, mark him as mixed-dominant as well, but much less than Plath. Interestingly, Hughes said that analyzing Shakespeare's complete opus nearly killed him and destroyed his immune system, which would imply that poetry came more naturally to him and analysis was a greater strain (Morrison 2001). Perhaps he was right. *Shakespeare and the Goddess of Complete Being* was published in 1992 and Hughes died in 1998 while undergoing treatment for colon cancer.

---

[1]   His archived books at Emory University include: Koestler's *Act of Creation*; Eliade; Gurdjieff; *The Cloud of Unknowing*; Shah, *The Wisdom of Idiots*; the *Upanishads*; J. Campbell; Gilgamesh; Weston's *Ritual to Romance*; Neumann, *The Great Mother*; Artaud, Bachelard, Blake, Borges, Coleridge, Gide, Graves, Pascal, Rimbaud, Shakespeare, Yeats; Castaneda, Crowley, John Dee, Jeane Dixon; Findhorn; *More Lives than One?*; witches; demonology; Jung's *Symbols of Transformation*; Neumann's *The Origins and History of Consciousness*; E. Jones, *Psycho-myth, Psycho-History*; R.D. Laing's *Self and Others*, *The Politics of Experience*; *Primal Scream*.

Genetic Predispositions and Childhoods

Plath's highly literate and hard-working parents expected academic success from their children. Her father, Otto Platt, had immigrated to the United States from Germany, changing his name to the variant Plath. Clearly a brilliant man, he learned English in one year and obtained a second Master's degree and a PhD from Harvard. Both a German instructor and a biology professor, he published scientific articles and a book on bumblebees.

Emotionally, he was less fit. An FBI file described him as having a 'morbid disposition' and possible Nazi leanings (Alberge 2012). His mother, a sister and a niece were all depressives. Sylvia Plath's son, Nicholas, would also commit suicide, implicating the genetic connection and/or early trauma from losing his own mother and a subsequent mother figure, Assia Weevil. Sylvia's trauma at losing her father to undiagnosed diabetes, gangrene, loss of his leg and untimely death when she was only 8 would have a long-lasting effect on her mental health and recurred obsessively in her poetry. To complicate matters, Nancy Hunter Steiner, a former roommate of Plath, reported the latter's irrationally guilty remark about her father's death: 'He was an autocrat... I adored and despised him, and I probably wished many times that he were dead. When he obliged me and died, I imagined that I had killed him' (Alexander 1991: 138).

Aurelia Plath was a bright and a tireless reader of literature. Born in an age even more repressive to women than her daughter's, she nonetheless managed to get a four-year degree allowing her to teach English, German and vocational subjects. She then worked on an MA in English and German and married her German professor. With his insistence, Aurelia stopped working and became pregnant immediately. A second child, Warren, arrived shortly thereafter. While Plath's father was reportedly selfish, domineering and anti-social, her mother possessed both a drive for success and a commitment to nurturing, perhaps overbearingly so. Aurelia Plath would dedicate herself to fostering her children's creative tendencies and aiding her husband's publishing efforts. She accentuated the importance of the written word, reading her young children nursery rhymes, fairy tales and poetry. She and her daughter would eventually exchange almost daily letters marked by a mutual love of literature and respect for each other's opinions. It should be noted, however, that Plath's abridged letters, edited by her mother, excised any talk of maternal negativity.

Hughes had a very different early childhood. When he was around 3 years old,[2] he followed his older brother Gerald into his 'North American

---

2   Although the age changed as Hughes aged, according to Middlebrook (2003).

Indian' fantasy of the 'Paleolithic' hunt. These years were formative to his mythopoeic imagination and he always harked back to them. Despite Hughes' contention that this was a paradisiacal period for him, handling dead animals at such a young age may well have traumatized him at an unconscious level, even while immersing him in the natural world with freedom to roam. Keith Sagar (2000/2006) reported that, following the future poet's last hunting expedition with his brother, his eyes were opened to the senseless pursuit of a fox, whose own eye he later tenderly closed in a dream. As Sagar wrote, Hughes' attitude toward hunting remained ambiguous. Rationally, he justified the hunt, but on an emotional (right-hemispheric) level, he remained closely identified with foxes and animals generally. Gerald, idolized by his brother and preferred by their mother, left home at 18. Ted, then, lost his brother when he was 8, the same age that Sylvia had lost her father.[3]

In the *Paris Review*, Hughes (1995) attributed his poetic prowess as an adult to his youthful hunting, which created 'a subsidiary' brain, nature-identified and body-based—a brain we might now interpret as right-hemisphere enhanced. A sense of melding with the hunted animal also suggests a lack of ego boundaries. As a child, he felt a fox, at the top of a hollow, 'had leapt into his head, supplanting his own provisional human nature with its own definitive foxhood' (Sagar 2000/2006: 42). Hughes eventually stopped hunting, claiming he had *viscerally* felt his wife's horror when he killed a wounded heather grouse (Middlebrook 2003: 68). He continued fishing, a sport he found both contemplative and a homecoming. His sense of returning to the true self when immersed in water also suggests an ecstatic, wordless, healing return to the maternal womb. In a way, his need to be in thrall to ever-renewed women, as we will see, reiterated that original merger with a divine mother/lover.

As to his early exposure to literature, Hughes' mother made up stories to entertain her young son. When he was 12, after his linguistic talent had been recognized at school, his mother bought him a children's encyclopedia of folktales. She later added 'a shelf of the classics, including the complete works of Shakespeare' (Middlebrook 2003: 67). His father's literary influence was confined to reciting Longfellow's 'Hiawatha'. The poem's rhythm and meter were recouped in his son's grade school recitations and creations (Middlebrook 2003: 71).

Hughes (1995) said he could 'never escape the impression' that West Yorkshire, England, where he grew up in a working-class home, was 'still in mourning for the First World War'. His father suffered PTSD, resulting in long-lasting silent depression. His mother claimed psychic powers he

---

3   Age eight curiously recurs as the time when poets lost an important figure in their lives.

believed she had transmitted to him. However, her eerie experiences most likely resulted from her own traumas: a brother had committed suicide (she said she felt his pain when the act was done); her dead sister appeared to her as an angel, signaling the coming death of family members (Middlebrook 2003: 68). Hughes also claimed he accessed occult information in dreams that had the troubling habit of coming true like his mother's visions. In his poem 'Source', he suggested his mother's sudden tears in his childhood might have flowed from future woes she was intuiting. While possibly true, a more likely source would have been the endemic sorrow brought on by war, deaths and her husband's depressive state. Yet, similarly, in *Shakespeare and the Goddess of Complete Being,* Hughes claimed that a poet's future self could dictate to him in the present.

Hughes would cultivate the paranormal dimension in hypnotic processes as well as in his dreams. After marrying, Plath learned dissociative techniques from her husband and also fell prey to the bizarre atmosphere in the Hughes family home. She sensed 'feelings of jealousy and forces of witchcraft and black magic', whether real or imagined (Alexander 1991: 197). Yet, Plath's early interests, especially Jungian, show she had already been fascinated by the kinds of activities and theoretical readings that Hughes would eventually advocate (see Christodoules 2008).

Early Poetic Inclinations

Hughes began writing poetry when he was 7, encouraged by a female teacher in his school. Plath had begun even earlier, at 5. Prior to Otto Plath's death on 5 November 1940, Sylvia and her brother had lived upstairs so as not to disturb him, descending nightly for a brief ritual of singing a song, listing names of insects or reciting a poem (Alexander 1991: 28). On 11 August 1941, the precocious Sylvia, now 8, had published her first poem in the Boston *Herald*. On the beach at 10 years old, Sylvia listened as her mother read Matthew Arnold's highly rhythmic poem, 'The Forsaken Merman' (Alexander 1991: 27). In this sad tale, a merman's Earthbound wife abandons him as he and their children return to the sea. The young and vulnerable Sylvia was struck hard, perhaps linking the merman to the loss of her father. She spontaneously composed her own poem on a subsequent night at the beach, while gazing up at the moon. The 10 year-old's intense emotional reaction to a poem, sparking her own creativity, again suggests enhanced right-hemispheric functioning. The young Sylvia and Ted both had deft ears for the stress, melody and meter of poetry, all right-hemispheric features of emotional prosody (Kane 2004). Additionally, the trauma of losing her father so young, along with her genetic predisposition to depression, may have caused dysregulation in the right hemisphere with subsequent mental problems (Schore 1994; Platt 2007).

Plath's ECT and Empathic Merging

After a stellar performance in high school encouraged by her English teacher, Wilbury Crockett, Plath went on to Smith College. Further accomplishments there lead to awards and published work and an internship at *Mademoiselle* magazine in New York City during the summer of 1953. As the summer concluded, while riding home from the train station with her mother and grandmother, Plath learned that she had not been accepted to a summer short story writing class at Harvard. This singular rejection, after her highly stressful summer, plunged her into a depression with an inability to read or concentrate. In a merged, desperate state, she would say: 'Oh, Mother, the world is so rotten! I want to die! Let's die together' (in Alexander 1991: 118).

Seeking medical advice, Aurelia Plath agreed to a series of electro-convulsive treatments (ECT) for her daughter as an outpatient. Badly administered, the treatments left Sylvia fearful they would be repeated and cripple her ability to write creatively again (see Ferretter 2008). She attempted suicide with sleeping pills; but, taking too many, she vomited and survived. 'Oh, no,' 'It was my last act of love', she weakly intoned in a 'completely coherent and rational voice', according to her mother, when found under the crawl space of their home (Plath 1975: 125-6).

On 28 December 1953, Plath wrote a letter she never mailed to 'E.', probably Eddie Cohen, a friend and correspondent at the time. She detailed the events following her *Mademoiselle* summer, including her perceived verbal failings: 'It turned out that not only was I totally unable to learn a squiggle of shorthand, but I also had not a damn thing to say in the literary world; because I was sterile, empty, unlived, unwise, and UNREAD. And the more I tried to remedy the situation, the more I became unable to comprehend ONE WORD of our fair old language'. She needed 'someone to love me, to be with me at night when I wake up in shuddering horror and fear of the cement tunnels down to the shock room, to comfort me with an assurance that no psychiatrist can quite manage to convey'. She confessed her 'hatred toward the people who would not let me die, but insisted rather in dragging me back into the hell of sordid and meaningless existence' (Plath 1975: 130-2).

In a letter dated 29 November 1956, addressed to 'Dearest Mother', Plath recalled that at the time of her suicide attempt she was 'terrified that if I wasn't successful writing, no one would find me interesting or valuable' (Plath 1975: 133). After the completed suicide in 1963, her mother theorized that 'Sylvia's tragic flaw lay in her own very weak ego strength' (quoted in Alexander 1995: 333, as an 'unpublished comment by Aurelia Plath').

Plath's first shock treatments are significant, not only for the traumatic and enduring effect they had on her, but also for what they say about her brain organization. At Guy's Hospital in London, researchers Fleminger and Bunce (1975), interested in cerebral dominance in left- versus right-handers, analyzed a pre-existing study on the effect of *one-sided* placement of ECT on verbal memory. These results had shown impaired verbal memory after left-sided placement, but not after right-hemispheric treatment. In their own study, they examined left-, right-, *as well as* mixed-handers. They found that mixed-handers were more likely than strong left-handers to be right dominant and emphasized the need for testing right-handed patients for language dominance to treat the appropriate side of the brain.

If, as described in her semi-autobiographical novel *The Bell Jar*, Plath had been treated on both sides simultaneously, she may have suffered increased verbal loss, especially if her language dominance were right or bilateral. From an emotional vantage, the results could have been even more severe. In recent research, we find that anxiety breaks down into two types: anxiety about the future expressed as verbal rumination in left frontal regions, versus panic disorder emanating from high arousal in the temporo-parietal region of the right hemisphere (Heller *et al.* 2008, and Smeets *et al.* 1997). Plath experienced verbal loss and insomnia, fear about her future and fear of abandonment—a lethal and recurring combination (Alexander 1991: 120). Her beloved English teacher, Mr. Crockett, pain-stakingly re-taught her the alphabet and spelling at McLean hospital, where she had been transferred, thanks to her benefactor, Olive Higgins Prouty. Plath could read and write again only after a month's time. Her mother noticed that the mood swings were more, not less, pronounced after ECT.

However, with complete faith in her new psychiatrist, Dr. Ruth Beuscher, Plath underwent a few more properly administered electroshock treatments for her depression that proved successful. The doctor attributed this speedy recovery to her patient's trust in her *and* to a need to be punished. Interestingly, new research also shows that 'brain scans of depressed patients receiving placebo treatment showed neurological improvements in certain parts of the brain that were identical to those seen in depressed patients receiving cognitive therapy or antidepressants. The mere *thought* of receiving proper treatment triggered a clear objective change in brain function among the depressed patients in the placebo effect group' (van Lommel 2010: 199). Plath's psychiatrist also gave her patient permission to hate her mother, displacing the pain and shame of her own self-loathing and/or misplaced guilt over her father's death.

In therapy, Plath used Jungian theory to explain her mother's failures in raising her. She had 'raised [her] with this intense focus on the thinking function, on intellectual performance, which was not her nature; she was an intuitive, feeling type'; she had not encouraged her to 'use a more affective part' of herself (in Alexander 1991: 130). Using a theoretical under-pinning that accentuated the neglected intuitive and feeling (right-hemi-spheric) functions both intensified her anger toward her mother and gave her solace. A journal entry from this time resounds with maternal hatred. Focusing on the 'bad' mother, Plath would even express anger at her therapist, now the 'good' mother, for having changed an appointment: 'She does it and is symbolically withholding herself, breaking a "promise", like Mother not loving me, breaking her "promise" of being a loving mother each time I speak to her or talk to her' (Plath 2000: 444). In a belated inter-view with Karen Maroda (2004), Beuscher admitted that she had focused on Plath's feelings toward her mother.

Nonetheless, Aurelia Plath believed herself to be a devoted mother who toiled in hardship to raise her children after her husband's death. There is no hard evidence to deny this, other than her semi-fictional portrayal in *The Bell Jar*. The UK edition, which Sylvia 'belted out' after ten years' trepida-tion, appeared in January 1963, one month before her completed suicide. According to a recent interview with Elizabeth Sigmund, a close friend at the time of Plath's death, Plath would never have wanted her novel pub-lished in the UK under her own name (at first it was not; later it was) while her mother was still alive (Jordison 2013). In 1970, Aurelia did object to the novel's US publication on grounds that uncamouflaged and inaccurate character portrayals of living persons, including her, would hurt them.

Returning prematurely to school in the spring semester of 1954, Plath resumed dating, but seemed unstable and cynical towards others, accord-ing to her mother. An experience on a trip to New York provides further evidence of Plath's right-hemispheric imbalance. As described to her mother in a letter, she saw a silent French film at the Museum of Modern Art identified as the 'Temptation of Saint Joan'.[4] The film concluded, Plath cried, purging herself of 'the buildup of unbelievable tension, then the release, as of the soul of Joan at the stake' (Plath 1975: 135). It is not insignificant that reading faces, especially with sad expressions, is in the domain of the right hemisphere. Plath most likely felt a synchronous connection between her own plight and the iconic female predecessor. Similarly, she would extol Cleopatra's suicide in her last poem, 'Edge', a few days before her own.[5] Severe stress can generate personally meaning-ful connections with mythic dimensions, as we saw with Jung, and these connections occur in an overactive right hemisphere (Beitman and Shaw

---

4   Probably, *La Passion de Jeanne d'Arc* (1925).
5   Kroll (1976) uncovered the Cleopatra connection and also pointed out Graves' identifying her as an incarnation of the Moon-goddess (151-4).

2009). Similarly, Plath's empathic feelings manifested *physically* in an instance where her arms blistered with welts at the exact time she *thought* Ethel and Julian Rosenberg were being executed (in Alexander 1991: 113). Delusions about the body and a confusion of self and other originate in the right temporo-parietal junction (Decety and Lamm 2007).

Hughes: Destiny, Dream Precognition and the Occult

At 19, Hughes had composed the poem 'Song', heard as a voice in the air, while on night duty in the RAF. Each stanza of 'Song' begins with the apostrophe 'O Lady', a mythic moon goddess represented in the wind and water, who inspires the poet. Heartbroken by the existence of her other lovers, the poet's head disintegrates in his hands, turning to dust as naturally as 'the world's decay'. While seemingly evoking Robert Graves' *White Goddess*, which would entrance Hughes as well as Plath, he did not receive a copy until 1951 when he went up to Cambridge, marking this dissociative poem as a possible precognition.

At Cambridge, Hughes had a highly significant dream, also bearing on his destiny, in which a burnt fox walked upright into the room, placing a human hand on the English paper he was begrudgingly writing. Leaving a bloody print, the fox said, 'You have to stop this. You're destroying both of us'.[6] Combining imagery of the fire heating his room with his totem animal, Hughes' unconscious jolted him into consciousness. This part of him, associated with nature, the hunt and his shamanistic self, initiated his decision to read archeology and anthropology rather than English literature.

On 25 February 1956, Hughes and Plath famously met at a party. In 'St. Boltophe's', a poem written many years later, Hughes affirmed their destiny: 'That day the solar system married us / Whether we knew it or not' (Hughes 1998: 14). Their astrological charts conjoined them, as surely as the 'ring-moat of tooth-marks' she branded on his face that night (Hughes 1998: 15). The bite had been her response to his grabbing her earrings and headband. But perhaps Plath too felt lured by destiny, already cognizant that 'one day I'll have my death of him', as she wrote later that night in her violent, erotically charged 'Pursuit' (Plath 1960/2008: 22). Plath sent her mother a copy of the poem for submission to a magazine. Whether habit or hypergraphia brought on by loss and trauma, her abundant letters home continued until a week before her suicide. Hughes, a prolific letter writer himself, sometimes divulged information that countered what he had written in published poems, as we will see.

---

6   See 'A Tribute to the Poet Ted Hughes': http://www.youtube.com/watch?v=18DdJO9Lg-s, where Hughes recounts the dream along with dramatic imagery.

Hughes also claimed an early dream in which an angel had shown him a small square of satin, later believed to be the same material lying under his dead wife's head in her casket.[7] This dream precognition resembled his mother's that announced an impending death in the family. During their marriage, the couple read Tarot cards. They also successfully worked a homemade Ouija board with an overturned glass, accessing 'spirits', hoping to equal Yeats and his wife in paranormal prowess (Alexander 1991). As with James Merrill and David Jackson or Victor Hugo's wife and son, it takes two to make that happen, suggesting a right-hemisphere to right-hemisphere connection. In fact, Hughes (2010) claimed he and his wife had a 'sympathetic', 'telepathic union' in a 'single shared psyche'. In the *Paris Review* (1995), he qualified this mind-merging as 'intrusive'.

Hughes claimed another important dream in 1956, the night before their wedding: 'he had hooked a pike and, when the fish began to surface, its head filled the entire lake', a possible omen of engulfment (Koren and Negev 2006: 88). In his American fishing magazine interview, Hughes called fishing 'a form of meditation, some form of communion with levels of yourself that are deeper than the ordinary self' (Hughes 1999: 50). But, in recurrent dreams throughout his life, Hughes' pikes symbolized more than his feelings in general: they also represented his relationship to women.

Assia Wevill, the woman who came between Plath and Hughes, recounted her own pike dream during a visit to the couple's Court Green home with her husband, David. In *Birthday Letters*, Hughes wrote about Assia, saying, 'After a single night under our roof / She told her dream. A giant fish, a pike / Had a globed, golden eye, and in that eye / A throbbing human foetus'.[8] Hughes added, she had 'sniffed' out the couple, mere 'puppets' for her Fates' 'performance'. She was a 'Black Forest wolf, a witch's daughter / Out of Grimm'; the 'Lilith of abortions'. Mind-to-mind contact had supposedly initiated this unwitting romance: 'I saw / the dreamer in her / Had fallen in love with me and she did not know it. / That moment the dreamer in me / Fell in love with her, and I knew it' (Hughes 1998: 157–8).

Interestingly, before their dream encounter, Hughes had published the poem 'Pike' in *Lupercal*, alluding to the visible eye of a pike that had been swallowed past its gills by another. Had Assia read 'Pike' and pre-cognitively dreamed of her own demise, with their coming child doomed

---

[7] See 'Ballad from a Fairy Tale' for the poetic rendering of this event (Hughes 2003: 171–2).

[8] Sagar (2000) reported: 'In a message read for him at the award ceremony of the Foreword Poetry Prize, Hughes said that in writing *Birthday Letters* over about 25 years he had "tried to open a direct, private, inner contact with my first wife, not thinking to make a poem, thinking mainly to evoke her presence to myself and feel her there listening"' (83).

to die along with her? Had the lovers' minds really become conjoined that night? Or had Hughes invented the dream details to still his dead wife's spirit and assuage his conscience? David Wevill found it unlikely his wife could have created such mythic imagery in her otherwise unexceptional dream life and did not recall her sharing the dream at all that day (see Koren and Negev 2006: 88–9).

Plath may well have been as intuitive as she claimed, immediately sensing this attraction. Her suspicions were confirmed after a camouflaged telephone call from Assia to their home. A bonfire of her husband's letters, papers and her second novel based on their loving marriage demonstrated the extent of her anger. On a prior occasion, a phone call from BBC producer Moira Dolan, in conjunction with his late return from a meeting with her, precipitated a similar shredding spree of his works in progress along with his precious copy of Shakespeare's plays.[9]

Different Neural Underpinnings

While mind melding and collaboration, sometimes fraught, were parts of their creative repertoire, Plath and Hughes had different types of minds from a gender perspective. Previc (2009)[10] said the male mind tends toward left-hemispheric dominance with its abundant dopaminergic connections and a 'detached', 'exploitative', 'linear' orientation towards future time and distant space. The female mind is more right-hemisphere dominant and serotonergic, interested in maintaining close, communal and empathetic ties. One could evoke the Paleolithic hunter versus the nurturing mother model. Even if Previc's theory does not address the particular case of poets with atypically organized brains, the basic paradigm works in describing the doomed collaboration of these two great poets.

Hughes possessed many traits of Previc's highly dopaminergic male. He was a hunter, with disheveled, unwashed ways. He was both highly intelligent and hypersexual, seemingly more interested in sexual conquest than in maintaining close bonds. He never recognized Assia Wevill's child, Shura, as his own. His belatedly discovered 'Last Letter' says he was with yet another woman, Susan Alliston, the weekend that Plath died. In the poem, he admits hiding from his emotionally fragile wife who had previously sent him a suicide note and then burned the letter in front of him when he rushed to see her. He abandoned Assia as well, reportedly maintaining a relationship with Brenda Hedden and other women, after marry-

---

9   Accounts of these events differ so it is hard to sort out the facts (see Bundtzen, 1998).

10   Serotonin inhibits dopamine and the serotonin system is 'predominantly localized in the right hemisphere' (Previc 2009: 39). This makes sense because right-hemispheric closeness is necessary for maternal bonding.

ing his devoted second wife, Carol Orchard (Koren and Negev 2006).[11]
Hedden, who broke off the relationship, plainly described Hughes' need
for conquest, not domesticity:

> He was a real hunter. The moment I drew away from him and became
> independent, I was more attractive in his eyes, and he chased me and
> pleaded that I would come back. It was the same with Assia: when she tried
> to break away and was out of his reach, he became motivated. But, when
> they were together, he did terrible things. I feared I would end up like her,
> and resisted his temptations. Her terrible suicide saved my life. (Koren and
> Negev 2006: 221)

Plath was highly intelligent, ambitious and sexually motivated as well, her
dopaminergic side. But, living in a time when women were labeled 'sluts' if
they acquiesced to male sexual overtures, she was conflicted. Calling her
first experience a rape, she went to hospital for hemorrhaging, yet continued dating the man. There were more men, sometimes involving consensual violent sex. Utterly taken with each other, Plath and Hughes
married only months after meeting. In her journals, she regularly mentioned their 'good' love-making sessions and told her mother when she
met Hughes that he was 'the strongest man in the world… with a voice like
the thunder of God' (Plath 1975: 233).

Clearly, Plath was attracted to Hughes for his raw strength as much as
for his poetry, having long avowed her need to be dominated by a strong
male who would not be jealous of her own poetic ambitions. She was also
aware that this need might be an overcompensation for her father's loss.
On the serotonergic side, Plath maintained a tidy home — proudly cooking
delicious meals and nurturing her children — and felt deeply bound to her
husband. In fact, during her marriage, as her serotonergic self waxed, her
creativity waned.

Mythic Underpinnings

Reducing the story of Plath and Hughes to their biochemistry would be
reductive, to say the least. The fragile, fatherless daughter suffered alternating emotional states, adopting theoretical constructs to support her
shaky sense of self. Judith Kroll (1976), who attended Smith College after
Plath, had gone to England for a summer but never met the poet. When she
later returned, Plath was already dead. *Ariel* had had a 'visceral' impact on

---

[11]   Koren and Negev (2006) also cite Hughes' letter to his brother Gerald in
which he 'confessed that he had finally found it impossible to stay married
to Sylvia, especially because of her "particular death-ray quality"', and that
he was pleased to have left her (110). Hughes also once compared himself to
a greylag goose, a bird that remains faithful to its first mate, in a letter to
writer and future lover, Emma Tennant (Koren and Negev 2006: 221).

her, so she decided to write her dissertation on Plath, later published as *Chapters in a Mythology* (1976/2007).

Kroll met with Hughes and his sister Olwyn, who hoped to sell Plath's papers to Vassar College. When Kroll identified the mythic underpinnings of Plath's poetry, Hughes called her 'clairvoyant'. But she had also worked hard to uncover Plath's deep admiration for and use of the theories of Graves, Frazer, William James and Jung, along with the poetic language of Yeats, Eliot, Lawrence and Woolf. Kroll now had rare access to Plath's creative process through the poet's papers and library of annotated books as well as Hughes' personal recollections. She also shared Plath's interest in 'the Tarot, séances, spirits' and 'Moon mythology', having already worked on a study of Yeats's *A Vision* (Kroll 1976/2007: xiii, fn. 3).

Plath had thoroughly integrated the language of the admired theorists and poets into her poetry, according to Kroll. The fiery *Ariel* poems deeply identified with Graves' mythic formula where the White Goddess, incarnating the waxing and waning moon, mediates between the dying and reviving gods. Likewise, Plath now conflated her overly- revered dead father with the overly-loved unfaithful husband, seeking to be done with both through her poetry. Kroll says, using Plath's words, the poet had become 'Lady Lazarus, the lioness, and the queen-bee', a lethal female figure in a 'self-sufficient maternal universe' (Kroll 1976/2007: 11). Her newly creative self would kill off the oppressive male, extol and nurture her children, and write her best poetry. Plath's anima rival, Assia Wevill, had set this transformation in motion as she replaced the too domestic, less inspiring wife.

In his *Oxford Addresses on Poetry*, which both Hughes and Plath had read, Graves (1961) distinguished Apollonian and Dionysian poetry, saying the Apollonian was composed in the 'forepart' of the mind while Dionysian, Muse-inspired poetry came in trance, 'at the back of the mind' (in Kroll 1976/2007: 229, fn. 25). Kroll recognized that Plath's style had already changed from a 'mannered, cerebral' style to an opening to the Dionysian, coinciding with Graves' lectures delivered earlier when Plath was a Fulbright scholar at Cambridge. Kroll suggested that the Gravesian influence, plus the hypnotic exercises that Hughes had taught her, fostered the major shift to the Dionysian in her final year of life:

> A common procedure, according to Ted Hughes in a conversation of June 1974, was to concentrate on a chosen topic, exploring its associations for a fixed length of time, to 'hand it over' to a speaker or persona, and then to put it out of mind until some predetermined time, usually the next day, when she would sit down to write on the subject. This technique combined a deliberate preliminary elaboration of detail with spontaneity in the actual writing of the poem, relying on the unconscious to select and organize the material. The procedure was, in effect, a deliberate stoking of the uncon- scious. It would also train her in letting another personality take over. In the

> late poems, that personality might be said to be her true self, *reigning
> supralogically* [emphasis added]. (Kroll 1976/2007: xxxix)

Graves had used these emphasized words in a lecture to describe how
'intuitive thought' arises along with a 'personal rhythm' in emotionally
driven Dionysian poetry. Whether defined as conscious/unconscious,
front/back or left/right, both Graves and Hughes seemed to be describing
a dissociative takeover.

In Plath's poem, 'Getting There' (6 November 1962), with its unrelenting rhythm and saturation in horrific sights, sounds and sensations, she
loses the Nazi identification of 'Daddy', but finds herself in a hellish nightmare. On a churning, 'screaming' train ride through war-torn Russia, she
encounters death, mud and mangled bodies at the train stop, accompanied
by the sound of thunder and guns. But, the rocking train carriages become
cradles and 'stepping from this skin / Of old bandages, boredoms, old
faces', she is freed from her baggage in a stripping process that brings to
mind Jung's NDE. But, unlike Jung, who was jolted upward into outer
space, the grounded Plath steps out from the 'black car of Lethe', '[p]ure as
a baby' to an unnamed 'you' (Plath 1960/2008: 247–9).

Kroll interpreted this ending as a vision of rebirth and an erotic over-
ture to the poet's husband. The 'baby' could well be a metaphor for
incorruptible purity, of a self washed clean of the many horrors of her past;
but there would be no return to her husband. Death would come and the
final poems were impregnated with it. As Hughes wrote to Plath's mother,
one month after the suicide, his actions were *madness*; his wife's reactions
were a kind of *madness*, 'when all she wanted to say simply was that if I
didn't go back to her she could not live' (15 March 1963: Hughes 2007: 215).
The trajectory of their marriage was clear in his own poetry. In 'Flounders',
they had been a 'single animal, a single soul', reminiscent of merged,
infantile, oceanic bliss (Hughes 1998: 63). In '9 Willow Street', they had
become 'Siamese-twinned, each of us festering / A unique soul-sepsis for
the other, / Each of us was the stake / Impaling the other' (Hughes 1998:
72).

Kroll postulated that Plath's last nine poems, up to the final 'Edge',
resolved Graves' mythic drama. With no way out of her troubles, leaving
behind her earthly body was not a defeat but a prelude to rebirth through
self-transcendence. While a worthy and comforting conjecture, one could
also surmise that Plath had found synchronicities between the Gravesian
mythic story and her own life for some time—a pattern that consoled and
empowered her. However, in dropping the mask of the mythic feminine,
she lost her defense against the chaos of her emotional storms. Her core
sense of self had disintegrated.

Raised a Unitarian, she had never been particularly religious. Yet, in the
weeks before her death she was saying, 'I am full of God', and 'I have seen

God. He keeps picking me up', according to Hughes (in Kroll 1976/2007: xxxi, fn. 18). Rather than rejoining her wild 'Mother', the Moon-Muse, or defeating her rival, she sensed a merger with the divine masculine, the Father god of her youth. In 'Mystic', written ten days before her death, she says, 'Once one has seen God, what is the remedy?' '[M]eaning' is everywhere, it 'leaks from the molecules' and Christ can be seen 'in the faces of rodents' (Plath 1960/2008: 268–9). This sense of being in direct communication with God and heightened religious significance in the smallest things are classic signs of over-charged temporal lobes in epileptics or mania in bipolars. Plath's history and present stress fit perfectly with Persinger's 'sense of presence', felt especially by women who are highly interested in creative writing or writers themselves, and use their adult intelligence to fulfill an infantile longing for the 'God-parent' who will 'remove the anxiety of existence' (Persinger 1987: 112).

She had tipped from depression into mania, with a new sense of religious zeal, entraining a euphoric feeling in the here and now. Plath's words and behavior on the night of the suicide, as recounted by her downstairs neighbor, Trevor Thomas, suggest a dopaminergic surge. Wanting to pay for stamps he had given her that night, she tells him, 'Oh! But I must pay you or I won't be right with my conscience before God, will I?' Ten minutes later, when Thomas finds her still in the hallway, he wants to call Dr. Horder, but Plath says, 'Oh, no please, don't do that. I'm just having a marvelous dream, a most wonderful vision' (in Alexander 1991: 329).

Despite the urgency and publishing potential of her fiery *Ariel* poems, as well as an upcoming assignment for the BBC and a slated meeting with her British editor that day, she committed suicide in the early hours of the morning. Fortunately, enough of her serotonergic self remained to protect her children in her self-aggressive act. The combination of serotonin hypofunction and impulsive acts of aggression is said to be heritable. Otto Plath's self-misdiagnosis, then refusal to treat his diabetes, could be considered a precedent for his daughter's suicidal tendencies. Research shows that widespread serotonergic abnormalities may account for depression *generally*; but serotonergic deficits *specifically* localized in the ventral prefrontal cortex, which controls impulsive activity, are associated with suicide when major precipitating factors like failure, loss of a job *or a relationship partner* are present (Dongiu and Patrick 2008).

Plath's mother wrote these poignant words regarding her daughter at the end of *Letters Home*: 'Her physical energies had been depleted by illness, anxiety and overwork, and although she had for so long managed to be gallant and equal to the life-experience, some darker day than usual had temporarily made it seem impossible to pursue' (Plath 1975: 500).

In *The White Goddess*, Graves said, 'True poetic practice implies a mind so miraculously attuned and illuminated that it can form words, by a chain of more-than-coincidences, into a living entity — a poem that goes about on its own (for centuries after the author's death, perhaps) affecting readers with its stored magic' (Graves 1948/1966: 490). But he was talking only about male poets inspired by a woman who incarnated the Goddess. Yet, Sylvia Plath lived a life seemingly ruled by 'more-than-coincidences', as though destiny itself had brought her to poetry, to Cambridge, to Hughes, to Graves, to Yeats's house to live and die. Her poetry lives on because of her intelligence, linguistic erudition and superior poetic prowess, whether cerebral or nearly autonomous. Her words are magical because they reflect the adored predecessors, but new imagery shouts or bemoans her own traumas, flowing to her own rhythm. Her beauty and maternal devotion played no small part in elevating her in the poetic pantheon. Caught in the maw of her own mythmaking, this poetic goddess died far too young.

Hughes' Lament

In *Birthday Letters*, composed over time but not published until the year he died, Hughes blamed Plath's death on an unholy female triune — her mother, her therapist and an unnamed 'manquée journalist' (probably Suzette Macedo) — for poisoning his wife's mind against him. In his late 'Howls & Whispers', he asked: 'What was poured in your ears / While you argued with death? / Your mother wrote: "Hit him in the purse." / …And from your analyst: / "Keep him out of your bed". / … / What did they plug into your ears / That had killed you by daylight on Monday?' But, in a letter to his sister Olwyn (February 1963), he accepted his own blame:

> On Monday morning, at about 6 a.m. Sylvia gassed herself. The funeral's in Heptonstall next Monday.
>
> She asked me for help, as she so often has. I was the only person who could have helped her, and the only person so jaded by her states & demands that I could not recognize when she really needed it. (Hughes 1998: 213)

One considerable blow had been the inscription Plath had found in a new copy of the red Oxford Shakespeare during a meeting in his flat. Weevil had replaced what Plath had jealously rent asunder.

One will never know what actually occurred between Sylvia and Ted during their last brief encounter on the evening of 8 February in Yeats's house; but, on returning to the home of her friend Jillian Becker, who had been watching the children, Plath appeared distinctly different from her formerly sobbing self, more angry than sorrowful, perhaps an emotional prelude to the self-aggressive act that sealed her fate.

Graves, Hughes and Right-Hemispheric Dissociation

As we saw, Graves first attributed 'true' poetry to trance-induced inspiration from the White Goddess. But he abhorred as 'charlatans or weaklings [those who] resort to automatic handwriting and spiritism' (Graves 1948/ 1966: 441). At the end of *The White Goddess*, he said outright: 'I am no mystic: I avoid participation in witchcraft, spiritualism, yoga, fortune- telling, automatic writing, and the like' (Graves 1948/1966: 488). In his fourth Clark lecture, he again described the necessity of inspired trance for poetic magic to occur, decrying the 'idle, foolish, Earth-bound spirits that hover around the planchette board or the pillow of sick men' (Graves 1955/1959: 124).

In his *Oxford Addresses*, Graves got more specific, identifying a front– back cerebral dichotomy for Dionysian poetry to triumph over the Apollonian. Discounting spiritualism, he did acclaim hallucinogenic mushrooms, in use since the earliest myths and religions, including the European Dionysus (Graves 1961/1962: 126–34). He even recounted his own experience taking psilocybin in an apartment overlooking the East River in New York. He called what happened 'schizophrenia', using the term of the day for a divided mind, while actually describing a dissociative takeover on a journey awash in sights, sounds and colors (Graves 1961/ 1962: 134–7). Graves' knowledge of ancient myths could only have enhanced those visions, which he was able to dismiss or bring on at will. Still, he differentiated his one drug experience as a *passive* trance that illuminated all his senses, from an *active* poetic trance, with 'pen running briskly across paper' (Graves 1961/1962: 140). Recent research indeed shows that psilocybin *decreases* activity in the brain, decoupling neural hubs that usually work together, which frees the mind, makes time irrelevant, and provides deep insights through a positive reframing of troubling past experiences (Carhart-Harris and MacLean 2012).

Keith Sagar (2011) was correct in pointing out that Hughes believed himself to have been the first to conceive of the idea of left and right hemi- spheric differences:

> By 1958 Hughes had independently come up with his own theory of the bicameral split. He had been looking at the Journals of the Psychical Research Society, and wrote to his sister Olwyn: 'I was surprised to find one theory stated & famously supported which I thought was my own—this was that in right-handed persons the left side of the brain has charge of all consciously-practised skills, capacities etc, but in the right lobe is the sub- conscious, or something deeper, a world of spirits. This is a half-baked sort of idea.'

> [*Sagar added*] It is hardly surprising that the idea was at this stage half-baked, since it was not until 1981 that Roger Sperry was awarded the Nobel Prize for his work on the split brain. Nevertheless, Hughes' idea is remarkably prophetic. (Sagar 2011: 7)

Indeed, Hughes had discovered that the Psychical Research Society's founder, F.W.H. Myers (1903/1954), had said 'dextro-cerebrality', i.e. right- hemispheric dominance, was more likely to produce 'spirit' contact and telepathic information than typical left-dominance. Hughes' interest in astrology, his attempts at telepathy and Ouija board communications, testify both to his beliefs and to his own enhanced right-dominance.

In the same letter cited by Sagar, Hughes explained how he had *intuited* the split mind from the eyes alone.

> ...many people's right eye has an utterly different expression than their left, a different brilliance etc. One of Dostoevsky's pupils was so completely expanded that no iris showed. The whole side of the face is different, but the effect is most noticeable in the eye. This made me think that the side of the brain corresponding to the funny eye must be in a different state of excitement than the other, and have different concerns. (Letter to Olwyn Hughes, March 1958: 123)

Anna Snitkina's book, *Dostoevsky Portrayed by his Wife: The Diary and Reminiscences of Mme. Dostoevsky* (1926), described her husband's eyes in a very similar way: 'I was struck by his eyes, they were different: one was dark brown; in the other, the pupil was so big that you could not see its color'. Hughes did not connect Dostoevsky's differing eyes to his severe bouts of epilepsy. More modern research would also have shown Hughes that epileptic lesions in the left hemisphere often transfer language dominance to the right. Even without epilepsy, the dominant eye is often larger than the other.

In an April 1957 letter to his friend Daniel Huws, Hughes made a curious distinction involving nerve input from the ear versus the eye, yet it translates into the front versus the back brain idea of Graves. Hughes said that for some poets their creative 'demon' enters via the 'nerve from the ear' connecting to the 'oldest part of the brain', which 'controls & receives sensations from all the muscles & organs of the body'. An admired American poet, John Crowe Ransom, was, for Hughes, this type of poet. The 'nerve from the eye', on the other hand, connects to the more recent part of the brain 'responsible for abstract & constructive thought'. Wallace Stevens fell into this category for Hughes, who cleverly offered a third scenario: an 'uncomfortable & incomplete concert of the two', which suggests bilateral dominance. Gerard Manley Hopkins was his example. Both highly analytical and creative, Hughes' distinctions predated current uses of neuroaesthetics in literary criticism. Hughes himself favored the 'primitive' type of connectivity that values the body in nature, which would be the right hemisphere (Hughes 2007: 96–7).

Not surprisingly, Hughes' early letters show an iconoclast, with little interest in paid work other than his own creative writing, possibly supplemented through family farming. Like Crow Ransom, Hughes extolled agrarian living and recognized that human suffering was owing to our underlying duality. Hughes contrasted the divided mind of societal humans with the unitary mind of animals, which, not surprisingly, could only be regained in the altered states of hunters, poets and shamans. Reading his essay, 'Baboons and Neanderthals: A Rereading of the Inheritors' (in Golding 1986/1987), one gets the idea that Hughes would have preferred *Homo sapiens* had remained Neanderthals so as not to have lost the 'world of superior senses, superior intuition' to hyperactive rational intelligence (Hughes in Golding 1986/1987: 164).[12]

Hughes was on a mission to prove he was 'right'. He voraciously read other poets and philosophers seeking descriptive myths and dissociative practices to explain or access occult knowledge and creativity. Graves had said Shakespeare 'knew and feared' the White Goddess (Graves 1948/1966: 426). In 1957, Hughes embarked on a project to read all of Shakespeare, highlighting his steadfast affiliation with the bard's productions as well as his means. His long analysis, *Shakespeare and the Goddess of Complete Being*, was published in 1992. Hughes learned that Shakespeare's techniques derived from Hermetic Occult Neoplatonism, a mystical school of thought that used ritual magic to glean wisdom and clairvoyance from hallucinatory figures. The Shakespearean 'affable familiar', who brings dissociative knowledge and luck, recurs in Hughes' letters, with a humorous case in point:

> When I consider how my affable familiar has sabotaged my every attempt at a normal profession, and sabotaged my whole life while I constrain it so, and how very affable & magically helpful & luck-bringing it is when I entertain it & its inventions, its fantasticalia, its pretticisms and its infinite verballifications—then I think it would be the best & most sensible course to make a career of humouring it. If you humiliate your devils, they avenge themselves, by paralysing your outer efforts. They operate like minor Furies. (Letter to Lucas Myers, 22 July 1957, in Hughes 2007: 105)

Hughes now converted his front/back, eye/ear thesis to the left/right dichotomy with a vengeance. Categorizing and systematizing Shakespeare, he combined the mythic/hemispheric model he had divined for maximum effect. In his *Paris Review* interview, Hughes said that Shakespeare's work was 'a total self-examination and self-accusation, a total confession—very naked' (Hughes 1995: n. pag.). With much more known about the brain, he could bolster his argument considerably. Robert Ornstein had published *The Psychology of Consciousness*, an influential book that outlined the known right/left-hemispheric differences.[13] Michael Gazzaniga, the split-brain

---

12 See Sagar's (2010/2012) 'Ted Hughes and the Divided Brain' essay. Hughes also wrote a letter after Frieda was born in which he discussed at length the 'discoursive rational abstract mind' versus the emotional brain (Letter to Aurelia and Warren Plath, about 17 December 1960, in Hughes 2007: 174–5).

13 See Ornstein (1972/1996: 81–125). Ornstein's update of his own lateralization research, *The Right Mind: Making Sense of the Hemispheres*, did not come out until 1997.

researcher who had shared the Nobel Prize with Roger Sperry, had published four books on the subject. Springer and Deutsch's 1989 edition of *Left
Brain, Right Brain: Perspectives from Cognitive Neuroscience*, was also available. Even if Hughes had not read these books, he may have been made
aware of their contents through other sources.

In some sense, Hughes the analyst may have been reductive in his use
of the neuroscientific materials; but Hughes the poet expressed his ideas in
a way that scientists could never have done. Personifying the hemispheres
using active, descriptive language, he explained how the left grabs the
right hemisphere's gift of imagery to make a point:

> What happens then is poetic. 'At wit's end is God', says the proverb. The
> moment the left side is brought to a halt, in this way, the loyal old right lobe,
> unembittered by its owner's official neglect, leaps forward with a suggestion
> in its own language—an image. The left side grabs it with relief, and out it
> comes as a metaphor. That metaphor is a sudden flinging open of the door
> into the world of the right side, the world where the animal is not separated
> from either the spirit or the real world itself. (Hughes 1992: 159)

At the time, Hughes would not have known that poetry itself is in large
part right-hemispheric language (Kane 2004). Rather, he described the
poetic process as full cooperation between the verbal left and the imagistic
right, adopting the current thinking that the right hemisphere was
'virtually wordless' (Hughes 1992: 157). Hughes' playful, but skillful,
description of the creation of metaphor exaggerated the effect, sounding
more like Persinger's super-charged temporal lobes or Newburg's ecstasy
of oneness:

> The balanced and sudden perfect co-operation of both sides of the brain is a
> momentary restoration of 'perfect consciousness'—felt as a convulsive
> expansion of awareness, of heightened reality, of the real truth revealed, of
> obscure joy, of crowding, indefinite marvels, a sudden feeling of solidarity
> with existence, with oneself, with others, with all the possibilities of being—
> a momentary effect, which is the 'poetic effect'. (Hughes 1992: 158)

In the *Paris Review* interview, Hughes described a special poem's
arrival, like 'Song', mentioned above, as an unconscious uprising from the
depths *à la* Graves. But, he also says Goethe might 'pick up a transmission
from the other side of his mind'. Hughes downplayed automatic writing
for himself. He required 'real, effortless concentration' to the point of
blacking out the windows and wearing headphones. Nor did he shun
corrections. Graves too had advocated molding the material after the initial
influx of inspiration. Even in 1994, Hughes was extolling handwriting in
the creative process because of the right-hemispheric 'subtext':

> Maybe the crucial element in handwriting is that the hand is simultaneously
> drawing. I know I'm very conscious of hidden imagery in handwriting—a
> subtext of a rudimentary picture language. Perhaps that tends to enforce
> more cooperation from the other side of the brain. And perhaps that extra

load of right brain suggestions prompts a different succession of words and ideas. (Hughes 1995: n. pag.)

On the other hand, he was adamant that poetry accessed *completely* without effort must *not* be reworked:

> My experience with the things that arrive instantaneously is that I can't change them. They are finished. There is one particular poem, an often-anthologized piece that just came—'Hawk Roosting'. I simply wrote it out just as it appeared in front of me. Why? 'Poems get to the point where they are stronger than you are. They come up from some other depth and they find a place on the page. You can never find that depth again, that same kind of authority and voice'. (Hughes 1995: n. pag.)

His final word on poetry, in this long, brilliantly articulated interview, favors the voice of this 'bottom level', not the right side of the brain.

Sagar used McGilchrist's (2009) work on hemispheric distinctions to show Hughes' early understanding of the poetic process. A poem like 'Hawk Roosting' differentiates the mad, 'disconnected' left hemisphere's imperious, entitled grasp of the world from the sane, sensual, body-based inner world of 'Wodwo', where the poet stalks his prey in right-hemispheric merged wonder in his natural surroundings. While an apt poetic approach in line with Hughes' reasoning, a scientific one would require more nuance. Similarly, Roderick Tweedy's book has overly dichotomized the left and the right hemispheres.

Left/right-hemispheric differences also have to do with fine detail versus global perception, both of which are necessary for full functioning. Even among animals, we find handedness and footedness based on these distinctions. Great apes are typically right-handed and the parts of the brain used in human language, Broca's and Wernicke's area, are associated with tool use in chimpanzees. So, grasping with the hand was probably the precursor for 'grasping' an idea using language. Even birds use their feet preferentially: the right foot exerts strong force and the left uses fine manipulations. Birds show right-eye (LH) superiority for discriminating visual patterns, the left eye (RH) for broader spatial tasks (see Sommer and Kahn 2009). Birds can also sleep using only one hemisphere, while the other stays vigilant. Unaware of this recent research, Hughes could only have projected his own ideal of undivided animal consciousness. However, he did recognize the crow's powers of mind with the keen eye of a hunter (see Klein 2008), while accenting their voracious appetites and rapacious behavior in *Crow*.[14]

---

[14] This collection, his first after Plath's suicide, was dedicated to the memory of Assia Wevill and their daughter Shura. Wevill committed suicide in the same manner as her predecessor.

Hughes' focus on the brain, whether as a top/down or left/right phenomenon, was more on the order of a liberating life stance, an explanation of his own mind (commingled with Shakespeare's), and a means to access deep, unfettered, poetic language via an uncoupling. The formula worked so well for him that he continued to look for it everywhere, exploring its interpretive advantages. In his introduction to *A Choice of Coleridge's Verse*, Hughes hypothesized that Coleridge's nightmare of a frightening woman who tries to rip out the elder poet's right eye might be an 'attempt, by some supercharged autonomous center of split-off consciousness in the right hemisphere of the brain to remove by physical violence — by terrorist violence — the over-policing, over-discursive, censorious vigilance of the left hemisphere of the brain, that was denying it access to life' (Hughes 1996: fn. 58). Using his over-the-top metaphor, Hughes would not have known that both eyes are connected to both hemispheres: the left side of each eye connects to the right hemisphere and the right side to the left hemisphere. Nonetheless, we all have a dominant eye, which usually reflects our cerebral dominance.

While there is much one could say about Coleridge, Hughes, in his introduction, focused on his forerunner's dual consciousness, i.e. the division between his Christian Self and his 'Unleavened Self' and his Mother complex (Hughes 1996: 6). As the last of ten children, Coleridge could not compete with seven brothers for the 'exclusive love' of their mother. In addition, he had lost his father at 10, and was sent to a school for orphans, returning home infrequently.[15] The hole left by the harsh/absent mother created both intense longing and some of the most Terrible Mothers in Romantic poetry. Hughes mentioned Coleridge's 'logomania', but never his bipolar disorder. His interest lay in Coleridge's conflicted feelings and the surrogate parents he had found in Wordsworth and his sister Dorothy. In addition, Wordsworth's ability to bring information to Coleridge's awareness on a shared 'wavelength of excitement', overcoming the younger poet's 'self-censorship', may well have recalled Hughes' own telepathic relationship with Plath (Hughes 1996: 85).

For Hughes, The Nightmare-Life-in-Death in 'The Ancient Mariner' was the prototype for Coleridge's frightening dream figure, representing the terrible, clawing feminine onto which he projected his emotions, fears, desires and violent tendencies. All three visionary poems — 'The Ancient Mariner', 'Christabel' Part I and 'Kubla Khan' — arose from the Unleavened Self, which Hughes identified as a hypnotic, supernatural sub-personality of the right hemisphere. What titillated Hughes in these master works had alienated Wordsworth, who withdrew his support and dashed Coleridge's

---

[15] Actually, Coleridge was 8, the same age that Sylvia Plath lost her father and Hughes' brother had left home.

poetic ambitions. However, briefly, the wound of early maternal neglect had rendered Coleridge a shaman singing the Mother goddess's song.

Curiously, Graves fit the same paradigm of maternal neglect as had Coleridge. The last of ten children, Graves most likely deemed his mother a desired but elusive figure in his early childhood as well. She was also puritanical and disapproving of his ways (Seymour-Smith 1982/1988: 405–6). After marrying, Graves' wife Nancy, a high-charged feminist, converted him to her views about the male oppression of women. His long-time muse, Laura Riding, who replaced Nancy, thoroughly dominated him. The myth of the White Goddess mirrored his relationship with Riding: 'He loved her, even when she seemed to him to be cruel, mocking, vicious and capricious, as the embodiment of poetry. Hence his later equation of her (among others) with the cruel aspect of the White Goddess. And it must not be forgotten that, since he had invented her before she came on the scene, he positively encouraged her to behave as she did' (Seymour-Smith 1982/1988: 215).

The Gravesian myth was not only about true poetic inspiration, but also expressed the yearning for the Mother, both loved and feared, that we see in so many male poets who experienced some form of maternal lack. The paradigm, personal as well as mythic, is seen in the stories of Isis and Osiris, Cybele and Attis, Ishtar and Tammuz, Aphrodite (Venus) and Adonis. Whereas *true* matriarchies never existed, the need to overturn them reflects the existential longings and fears of men, even womb envy, and explains the misplaced misogyny that appears in so many myths, religions and archaic rituals. Ultimately, Graves' vision of the Goddess in her cruel aspect reduces too one-sidedly to the man-eating woman whom Coleridge had depicted. We are reminded of Plath's *Ariel* poems identifying with that same aspect: the destructive queen bee who wished to destroy the men who had failed her in life.

Hughes thoroughly took up the mythic theme in *Shakespeare and the Goddess of Complete Being*, as he attempted to order the bard's entire opus based on the portrayal of the mythic feminine in his plays. The book's cover art shows a wild boar goring Adonis with a mortal wound. Hughes believed, in line with Graves' mythic scenario, that a major trauma was needed to confer special knowledge on the poet. Shakespeare's trauma was the 'tragic error' of abandoning his wife in Stratford while he went to London, compounded at the societal level by the suppression of the Catholic mariocentric tradition in England. Shakespeare's 'visionary' or 'shamanic' poetry, Hughes hypothesized, erupted at the confluence of these major crises. Shakespeare too was double: 'mythic' and 'realist'; female and male; incarnating both 'the Goddess's suffering' and 'the Puritan that makes her suffer' (Hughes 1992: 62). He had two 'vatic personalities [that fought] to come to terms inside his head—and inside his heart and throughout his nervous system'.[16]

Hughes' hemispheric theory then goes full throttle. The archaic, right-hemispheric Goddess myth—matriarchal, emotional and body-based—is transformed into a new, left-hemispheric, Goddess-destroying myth, which is patriarchal, rational and idealized. The Female of the right is 'inseparable from the womb memory, infant memory, nervous system and the chemistry of the physical body, possessed by all the senses and limitless'; the Female of the left is 'Puritan... idealized, moralized and chaste' (Hughes 1992: 161). Hughes impressively describes the neuroscientific reality of the right hemisphere as we now know it: the right hemisphere is dominant in the womb until language skills are learned in early childhood; the right thalamus is the cerebral processing center for all sense modalities; the sense of a limitless self occupying all space is right hemispheric.

The Shakespearean Hero divides the Complete Being of the Female into opposing aspects—Sacred Bride/Divine Mother versus Queen of Hell. When the mythic feminine subdivided, what happened next was 'madness', the very term Hughes had used to describe his own flight from Plath in his letter of 13 May 1963 to her mother: 'my love for her simply underwent temporary imprisonment by something which can only be described as madness, as much an attempt to free myself from the strangling quality of our closeness as by any other cause. My love for her simply continues, I look on her as my wife and the only one I shall ever marry, and these two children are ours' (Hughes 2007: 218). These words were designed both to console Aurelia and to keep her from visiting the children. Hughes did remarry.

For Hughes, the 'mad' turnaround in Shakespeare, mirroring the overthrow of the real *and* the mythic mother, derived from the fact that *all* of life is doomed. Just as the growing boy must overthrow the 'possessive control of the Female', his mother, in order to become a man, the mythic hero must overthrow the Mother Goddess because of her 'magical, terrifying, reproductive powers', 'the occult power of her paralysing love', and the 'uncontrollable new sexual energy which is searching for union with the unknown Female' (Hughes 1992: 326–8). Nonetheless, a great shift in

---

16 Hughes denied Plath's moon goddess affiliation, saying she was caught between two Sun-God myths—the story of Phaeton, 'who takes the reins of his father's Sun-chariot, loses control and is wrecked)—and the story of Icarus "who flew too near the sun" and fell'. Both Sun-God myths suggest to Hughes Plath's 'inaccessible, worshipped father' and 'a strong visionary and mystical core (the core of her mythic personality)'. The Phaeton myth was behind 'Ariel' and Icarus in 'Sheep in Fog', although she was unaware of it. However, first written with a hopeful ending, 'Sheep in Fog' was consciously revised to reflect 'her own soul's story' (Hughes 1992: 40–2).

Shakespeare's plays coincided with the bard's loss of his own mother in 1608, along with the move to Blackfriars Theatre. From this point onwards, saving, rather than killing off, the Female becomes his credo. Restoration of the Divine Female heals the crime against her, so that it cannot occur again.

In writing so obsessively about Shakespeare, even if originally in preparation for his *Choice of Shakespeare's Verse*, Hughes seemed to be thinking profoundly about his own personal life and traumas. Shakespeare's work encapsulated the teetering predicament of every male seeking love without the 'tragic' loss of his independence.[17] In fixating on Shakespeare's underlying myth, spelling out the Female's doom and resuscitation, Hughes had found a twinning of his own sad tale, supported by his slow simmering, now fully developed, hemispheric thesis. Hughes' mad flight, which resembled Shakespeare's, had precipitated his wife's tragic suicide, the first in a series of deaths that would encompass his most intimate female relations: a wife; his lover, Assia; their daughter; and his own sick mother, who died shortly after learning of Assia's suicide. On the stage of Hughes' life, female corpses of Shakespearean proportion were piling up and the trauma ran deep. *Birthday Letters*, like Shakespeare's final efforts, was Hughes' attempt to restore his first wife and to disculpate himself.

Hughes' theory of the cerebral hemispheres and the two-sided mythic feminine dovetailed with his own creative and personal needs. At a very basic level, following the Gravesian equation, he needed the inspiration of a woman to work. Beginning with the impetus of his mother's stories, her books, her psychic gifts; next, the encouraging voice of a woman teacher at school; he would later seek, then abandon, successive females whose energetic infusion enflamed, then ceased to kindle, him. After marriage to his second wife, Carol Orchard, he continued searching for women, introducing one, Jill Barber, personally to Robert Graves as his muse.

The first draft of *Shakespeare and the Goddess of Complete Being* was written in the form of fifty-four long letters addressed to a woman friend, Donya Feuer. Hughes restored his mother in *The Remains of Elmet*, written in collaboration with a woman photographer, Fay Godwin. In 'The Offers' (*Howls and Whispers*), he dreams of his dead wife's three-fold return to him, incarnating the mythic theme that had somehow stitched together the elements of his married life to Plath. Even his 'brain's hemispheres' appear in this poem, 'twisted slightly out of phase / To know you you yet realize that you / Were not you. To see you you and yet / So brazenly continuing to be other'. The poem ends with Hughes stepping into a bath, the

---

17 This need to escape the swamp-like Maternal/Feminine in order to live as an independent, rational, creative male is fully consonant with Camille Paglia's thesis in *Sexual Personae*, focusing on the negative aspect of the Goddess, not her nurturing side.

vulnerable place where Graves said mythic men die. The voice of his fallen Goddess speaks clearly to him: 'This is the last. This one. This time / Don't fail me'.

Hughes' poetic vocation had been destined, willed and followed with almost spiritual devotion. His neurological/mythic model allowed a sometimes heartless predator, a paranormally inclined poet and a self-styled shaman to explain himself and excuse his flaws. In the process, he left a dense trail of words, both poetic and analytical, to mesmerize us all. Despite his intricately worked cover story – an archetypal inflation – he remained human. In 'Last Letter', discovered after his death, Hughes recorded his errors and the full meaning of his failings in Plath's last weekend.

***

Near the end of *Crow*, in the poem 'Lovesong',[18] a question was asked: 'Who paid most, him or her?' A. Alvarez, who knew Plath and Hughes equally well, offered a response extracted from his book and printed in *The Guardian*:

> I had always believed that genuine art was a risky business and artists experiment with new forms not in order to cause a sensation but because the old forms are no longer adequate for what they want to express. In other words, making it new in the way Sylvia did had almost nothing to do with technical experiment and almost everything to do with exploring her inner world – with going down into the cellars and confronting her demons. The bravery and curious artistic detachment with which she went about her task were astonishing – heartbreaking, too, when you remember how lonely she was. But when it was all over, I no longer believed that any poems, however good, were worth the price she paid. And I've sometimes wondered if all our rash chatter about art and risk and courage, and the way we turned rashness into a literary principle, hadn't egged her on. (Alvarez 1999)

Hughes had his own theory, speculating in his poem 'Ouija' that his wife's tears and fears about the prospect of being famous had produced 'some still small voice' he could not hear prophesizing the future:

> Fame will come. Fame especially for you.
> Fame cannot be avoided. And when it comes
> You will have paid for it with your happiness,
> Your husband and your life. (Hughes 1998: 56)

---

18 Koren and Negev reported that Hughes gave Brenda Hedden a necklace identical to one he had given Assia. He also gave Brenda a copy of 'Lovesong' he had already given to Assia, 'which was in fact intended for Sylvia' (Koren and Negev 2006: 189). All of his women were as one goddess for him.

Conclusion

I set out to write about collaborating literary couples who used dissociative practices to access creativity. All of the individuals exhibited outer signs of atypical lateralization and had experienced the loss of a parent or sibling at an early age. Most sought access to lost loved ones through spiritualism, making various paranormal claims in the process.

Whereas the general credo in current neuroscientific literature says we use the whole brain in everything we do, it is clear that hemispheric differences and relative dominance are highly significant and have been since the beginning of human history. From the shamans of the Paleolithic age, with their hyper-realistic artistic skills and ritual practices, to contemporary autistic savants exhibiting extraordinary skills, lateralization matters. Shamanism, neurodevelopmental or organic disorders, especially epilepsy and bipolar disorder, can similarly favor different forms of creative artistic expression.

In addition, sensed presences, poetry and religious ideation engage an enhanced right or synchronous bilateral activation of the hemispheres, according to many theorists. Two other hemispheric dichotomies — left positive/right negative, left approach/right avoidance — are fundamental to understanding how the mind works. While a too bilateral brain can predispose to mental disorder of the schizophrenic or bipolar variety, it can also produce epiphanies, healing revelations and dissociative others with messages worthy of consideration.

The common claim among the practitioners of various techniques, explored in Chapter 2, was the inability to distinguish between waking and dreaming, suggesting an indeterminate state where imaginative thought — the mythopoeic imagination — ruled. Loss of the fully conscious self invites dissociative interventions and enhanced performative skills. Pure perception seems to trump linguistic cognition for ways of knowing with healing benefits for self and other. Dissanayake suggested that numinous perception and the origins of art began in maternal–infant relational behavior. Oubré corroborated, saying rhythmic activities create positive emotional states to unite the group.

Because a child's mind is exposed to environmental pressures both inside the womb and in early childhood development, proper maternal care is crucial. In this study, our male poets had attachment issues with their mothers; others suffered separation or loss. For Plath, and for girls in general, a properly present father is equally important. Verbal stimulation, through voracious reading and writing, when highly significant, primes the brain emotionally to produce a dissociative sense of presence dictating novel thoughts, by actually recombining what is already known, in a creative way.

While the therapeutic process can certainly heal, it must be cautioned that Freud, Jung, Flournoy and others pushed patients in a trance state into further dissociative splits through suggestive questioning. Decreased frontal lobe activity, as Newberg found, allows loose 'unwilled' associations to concatenate. Jung had encouraged his own cousin's reincarnation fantasies, as Flournoy did with Hélène Smith. Freud realized that his suggestions of improper sexual contact between his patients and their fathers in early childhood led to false admissions in his 'hysterical' patients. One can hardly separate the real from the imaginary in the many scenarios described in books by therapists making claims in Chapter 4. But there, too, it seems suggestion played a role in the therapeutic dyads, finding and perpetuating reincarnation fantasies in their patients, always revolving around sex and violence.

Despite these caveats, sometimes during highly stressful, relaxed or dream states, information slips into consciousness beyond the 'known' in any logical sense. My own dream experience pointing me to Jung's theories and Keats's poetry seemed farther afield than a suggestion or a prior reading. A trauma, an answer to an important question and a need to return to emotional homeostasis seem to stimulate these uprushes of dissociative words, with an unfettered right hemisphere facilitating the process.

Myers (1903) credited right-hemispheric dominance with a greater capacity for telepathy. Prescott (1922) thought that poetic language was visionary, transcending time and space. Fleck *et al.* (2008) connected magical thinking, creativity and belief in the paranormal to the right hemisphere. Carson (2011) showed how mental illness in very intelligent people allows for unusual perceptions and associations when escaping the over-systematizing left hemisphere's influence. Jung had so many uncanny experiences that he could not deny the existence of occult phenomena, explained as quantum physical events.

Other twenty-first-century therapists claim their patients can express knowledge drawn directly from them, as the two become mentally 'entangled'. Some alternative healers describe their intuitive healing practices as right-hemispheric or thinking aside. I might add that my friend who started me on this quest has become an intuitive healer herself. Neuro-scientists Fleck *et al.* have recorded increased right-hemispheric activity in study participants who claimed paranormal effects and had above average creativity. Schore and Bromberg believe that a right-hemisphere-to-right-hemisphere connection replicating a synchronous maternal–infant bond is required in the therapeutic process.

In this book, of the six poets studied, only Sylvia Plath and James Merrill underwent therapy. Neither uncovered memories of sexual abuse. Plath, who suffered from bipolar disorder, had lost her father when 8 years old. A therapist later focused excessively on Plath's relationship with her mother. Plath discovered Jung's theories on her own and applied them to her relationship with her mother as well. Retrieving the lost father in her husband, who then abandoned her, played a large role in her suicide. Previously, badly applied electroshock treatments and, in the end, inappro-priate drug therapy conspired against her. Merrill had once sought therapy for writer's block and recognized the emotional aftermath of his childhood traumas. He had lost a beloved governess at 10, followed by his parents' separation. Unresolved conflict with his mother over his sexual orientation and lack of procreation both plagued him and fueled his 560-page poem with a maternal surrogate and an intricate compensatory reincarnation theme.

While trauma need not be sexual to produce creative dissociation, gender relations, real or mythic, always seem to play a role. In Keats's 'Hyperion' poems, the Gods and Goddesses speak into the ear of their charges or penetrate their skulls through mental merger. Within the poems, dream characters materialize and a Goddess's silent face transfers know-ledge to the poet's brain in a panoramic flash. An NDE prefaces a Blakean opening of the doors of perception. Metaphors elucidate divine thought and poetry is born. Sometimes time and space are transcended. The tenor and pallor of Keatsian thought may suggest a precognitive sense that he would soon die.

Environmental conditions were also dissociative triggers. Hugo claimed the wind and waves spoke to him. He perceived sentient souls in flowers and dogs. The dead spoke to him in dreams and at the séance tables. Furthermore, he did not need to be present for the conjured spirits to speak his thoughts, using his metaphors. The 'spirits' had no specific sense of time and did not make prophetic warnings. However, wisdom proffered dissociatively seemed a prescient version of Talbot's holographic universe.

Rilke heard words in the wind and created poetry as if from dictation, perhaps precognitively sensing his imminent demise as well. He believed that poetic consciousness was collective and collaborative, without recog-nizing the role of genetic predisposition, trauma, voracious reading, per-sonal dramas and national crises. Yeats, with his panoply of predisposing factors, had the good luck (or destiny?) to marry a left-handed artist who had lost her father. The two mystical aspirants collaborated daily for two years in automatic handwriting or 'sleaps' to bring forth metaphors for poetry and a reincarnation scheme answering their questions and resolving their issues.

Hughes had a deep understanding of right-hemispheric processes, believing that he had discovered the split brain before the neuroscientists. But he also recognized a top/down dichotomy. He believed in astrology and had precognitive dreams like his mother. His lengthy study of Shakespeare pivoted on a matriarchal view of human consciousness and his own fear of female engulfment. The belief that he and his first wife were both shamans, and that the poet's future self could dictate to him in the present, expressed his own right-hemispheric proclivities. He had an extraordinary mind, as logical as it was analogical, as sophisticated as it was nature-bound. Unfortunately, his hunting instinct extended to lovers and contributed to the equally brilliant Plath's undoing.

Bibliography

Abrams, M.H. (1971) *Natural Supernaturalism: Tradition and Revolution in Romantic Literature*, New York: W.W. Norton & Company.

Ackroyd, P. (1995/1996) *Blake*, London: Minerva.

Alberge, D. (2012) FBI files on Sylvia Plath's father shed new light on poet, *The Guardian*, Friday 17 August, [Online].

Alexander, P. (1991) *Rough Magic: A Biography of Sylvia Plath*, New York and London: Viking.

Alvarez, A. (1999) How Black Magic Killed Sylvia Plath, *The Guardian*, Tuesday 14 September, [Online].

Ammari, E.H. (2011) Prodigious polyglot savants: The enigmatic adjoining of language acquisition and Morrison emaciated potentials, *International Journal of Business and Social Science*, 2 (7), pp. 158–173.

Andreasen, N. (2014) Secrets of the creative brain, *The Atlantic*, [Online], http://www.theatlantic.com/features/archive/2014/06/secrets-of-the-creative-brain/372299/ [accessed 25 June 2014].

Anthes, E. (2010) Ambidexterity and ADHD, *Scientific American Mind*, July/Aug, p. 7.

Anthes, E. (2010) Soothing traumatized children, *Scientific American Mind*, Nov/Dec, p. 9.

Armstrong, K. (1993) *A History of God: The 4,000-Year Quest of Judaism, Christianity and Islam*, New York: Ballantine Books.

Asai, T., *et al.* (2009) Schizotypal personality traits and atypical lateralization in motor and language functions, *Brain Cognition*, 71 (1), pp. 26–37.

Atmanspacher, H., Römer, H. & Walach, H. (2002) Weak quantum theory: Complementarity and entanglement in physics and beyond, *Foundations of Physics*, 32 (3), pp. 379–406.

Augustine (1992) *The Confessions of Augustine*, Chadwick, H. (trans.), Oxford and New York: Classics/Oxford UP.

Bachelard, G. (1942) *L'Eau et les rêves*, Paris: Librairie José Corti.

Bahrami, B. (2014) Finding an old flame, *The Pennsylvania Gazette*, Jan/Feb, pp. 31–35.

Baker, R.A. (1996) *Hidden Memories: Voices and Visions From Within*, New York: Prometheus Books.

Baradon, T. (ed.) (2010) *Relational Trauma in Infancy: Psychoanalytic Attachment and Neuropsychological Contributions to Parent–Infant Psychotherapy,* London and New York: Routledge.

Barks, C. (1997) *The Essential Rumi,* Edison, NJ: Castle Books.

Barnett, K.J. & Corballis, M.C. (2002) Ambidexterity and magical ideation, *Laterality,* 7 (1), pp. 75–84.

Barron, F. (1969) *Creative Person and Creative Process,* New York: Holt, Rinehart and Winston.

Baudouin, C. (1943) *Psychanalyse de Victor Hugo,* Genève: Editions du Mont-Blanc.

Bauer, M. (2003) *This Composite Voice: The Role of W.B. Yeats in James Merrill's Poetry. Studies in Major Literary Authors 24,* New York & London: Routledge.

Belluck, P. (2014) Study finds that brains with autism fail to trim synapses as they develop, *New York Times,* 21 August 2014, [Online, accessed 21 August 2014], n. pag.

Beitman, B.D. & Shaw, A. (2009) Synchroners, High Emotion, and Coincidence Interpretation, *PsychiatricAnnalsOnline.com.*

Berlin, H. & C. Koch (2009) Neuroscience meets psychoanalysis, *Scientific American Mind,* April, May, June, pp. 16–19.

Bloom, H. (1987/1989) *Ruin the Sacred Truths: Poetry and Belief from the Bible to the Present,* Cambridge, MA, and London: Harvard UP.

Bloom, H. (1997) *The Anxiety of Influence,* Oxford: Oxford UP.

Bolte Taylor, J. (1996) *My Stroke of Insight: A Brain Scientist's Personal Journey,* New York: Viking.

Bolte Taylor, J. (2008) *My Stroke of Insight TED Talk,* [Online], http://www.ted.com/talks/view/id/229, [accessed 29 August 2014].

Braude, S.E. (2003) *Immortal Remains: The Evidence for Life after Death,* Lanham, MD, Boulder, CO, New York, Toronto, Plymouth: Rowman & Littlefield.

Breton, A. (1964) *Nadja,* Paris: Gallimard.

Bright, K.J. (2013) Messages from the gods: Outsider art and the voices in Augustin Lesage's head, *Dangerous Minds,* 30 September, [Online] http://dangerousminds.net/comments/messages_from_the_gods_ outsider_art_and_the_voices_in_augustin_lesages_head, [accessed 1 November 2013].

Bromberg, P.M. (1998) *Standing in the Spaces: Essays on Clinical Process, Trauma, and Dissociation,* Hillsdale, NJ, and London: The Analytic Press.

Bromberg, P.M. (2006) *Awakening the Dreamer: Clinical Journeys,* Mahwah, NJ, and London: The Analytical Press.

Brown, M. (1997) *The Channeling Zone: American Spirituality in an Anxious Age,* Cambridge, MA: Harvard UP.

Bucke, R.M. (1961/1993) *Cosmic Consciousness: A Study in the Evolution of the Human Mind*, New York: Citadel Press/Carol Publishing Group.

Bundtzen, L.K. (1998) Poetic arson and Sylvia Plath's 'Burning the Letters', *Contemporary Literature*, 39 (3), pp. 434–451.

Burki, A., Cherkas, L., Spector, T. & Rahman, Q. (2011) Genetic and environmental influences on female sexual orientation, childhood gender typicality and adult gender identity, *PLOS ONE*, 6 (7), pp. 1–6.

Burnand, G. (2013) A right hemisphere safety backup at work: Hypotheses for deep hypnosis, post-traumatic stress disorder, and dissociation identity disorder, *Medical Hypotheses*, 81, pp. 383–388.

Butterfield, B. (no date) *The Troubled Life of Vincent Van Gogh*, [Online], http://bonniebutterfield.com/VincentVanGogh.htm, [accessed 17 September 2014].

Callan, E. (1975) W.B. Yeats's learned Theban: Oswald Spengler, *Journal of Modern Literature*, 4 (3), pp. 593–609.

Campos, D.J. (2011) *The Shaman & Ayahuasca*, Roman, A. (trans.), Overton, G. (ed.), Studio City, CA: Divine Arts.

Capra, F. (2007) *The Science of Leonardo*, New York, London, Toronto, Sydney, Auckland: Doubleday.

Carey, B. (2008) Study finds prior trauma raised children's 9/11 risk, *New York Times*, 5 February.

Carhart-Harris, R. (2012) How do psychedelics affect the brain to alter consciousness?, *Toward a Science of Consciousness Conference*, Tucson, AZ, 14 April.

Carreiras, M. & Price, C.J. (2008) Brain activation for consonants and vowels, *Cerebral Cortex*, 18 (7), pp. 1727–1735.

Carson, S.H., Peterson, J.B. & Higgins, D.M. (2003) Decreased latent inhibition is associated with increased creative achievement in high-functioning individuals, *Personality Processes and Individual Differences*, 85 (3), pp. 499–506.

Carson, S.H. (2011) Creativity and psychopathology: A shared vulnerability model, *La Revue canadienne de psychiatrie*, 56 (3), pp. 144–153.

Christodoulides, N. (2008) Sylvia Plath's 'The Magic Mirror': A Jungian alchemical reading, *Plath Profiles*, 1, pp. 247–258.

Claridge, G., Pryor, R. & Watkins, G. (1990) *Sounds from the Bell Jar: Ten Psychotic Authors*, Cambridge, MA: Malor Books.

Cohn, J. (2013) *The Minds of the Bible: Speculations on the Cultural Evolution of Human Consciousness*, revised ed., Henderson, NV: The Julian Jaynes Society.

Conforti, M. (1999) *Field, Form, and Fate: Patterns in Mind, Nature, and Psyche*, Woodstock, CT: Spring Publications, Inc.

Connors, K. & Bayley, S. (2007) *Eye Rhymes: Sylvia Plath's Art of the Visual*, Oxford: Oxford UP.

Cook, W.R. & Herzman, R.B. (2004) *St. Augustine's Confessions*, The Teaching Company: Audio/Visual Course.

Damasio, A. (1999) *The Feeling of What Happens: Body and Emotion in the Making of Consciousness*, San Diego, CA, New York, London: Harcourt.

Damasio, A. (2003) *Looking for Spinoza: Joy, Sorrow, and the Feeling Brain*, Orlando, FL, Austin, TX, New York, San Diego, CA, Toronto, and London: Harcourt.

Damasio, A. (2010) *Self Comes to Mind: Constructing the Conscious Brain*, New York: Pantheon Books.

Damrosch, L. (2005) *Jean-Jacques Rousseau: Restless Genius*, Boston, MA, and New York: Houghton Mifflin.

D'Aquili, E. & Newberg, A.B. (1999) *The Mystical Mind: Probing the Biology of Religious Experience*, Minneapolis, MN: Fortress Press.

Dawkins, R. (2003) *BBC Two*, Thursday 17 April.

Decety, J. & Lamm, C. (2007). The role of the right temporoparietal junction in social interaction: How low level computational processes contribute to meta-cognition, *The Neuroscientist*, 13, p. 6.

Devinsky, O. (2000) Right cerebral hemisphere dominance for a sense of corporeal and emotional self, *Epilepsy & Behavior*, 1 (1), pp. 60–73.

Dissanayake, E. (2000) *Art and Intimacy: How the Arts Began*, Seattle, WA, and London: A McLellan Book/University of Washington Press.

Ellenberger, H.F. (1970) *The Discovery of the Unconscious: The History and Evolution of Dynamic Psychiatry*, New York: Basic Books.

Ellmann, R. (1948/1979) *Yeats: The Man and the Masks*, New York and London: W.W. Norton & Company.

Epstein, G.N., MD (1981/1992) *Waking Dream Therapy: Unlocking the Secrets of Self Through Dreams & Imagination*, New York: ACMI Press.

Faurie, C. & Raymond, M. (2004) Handedness frequency over more than ten thousand years, *Proc. R. Soc. Lond. B (Suppl.)*, 271, S43–S45.

Ferretter, L. (2012) Just like the sort of drug a man would invent: The Bell Jar and the feminist critique of women's health care, *Plath Profiles*, pp. 136–158.

Flaherty, A.W. (2004) *The Midnight Disease: The Drive to Write, Writer's Block, and the Creative Brain*, Boston, MA: Houghton Mifflin.

Fleck, J.L., Green, D.L., Stevenson, J.L., Payne, L., Bowden, E.M., Jung-Beeman, M. & Kounios, J. (2008) The transliminal brain at rest: Baseline EEG, unusual experiences, and access to unconscious mental activity, *Cortex*, 44, pp. 1353–1363.

Fleminger, J.J. & Bunce, L. (1975) Investigation of cerebral dominance in 'left-handers' and 'right-handers' using unilateral electroconvulsive therapy, *Journal of Neurology, Neurosurgery, and Psychiatry*, 38, pp. 541–545.

Flournoy, T. (1899/1994) *From India to the Planet Mars: A Case of Multiple Personality with Imaginary Languages,* Shamdasani, S. (ed.), Foreword Jung, C.G., Commentary Cifali, M., Princeton, NJ: Princeton UP.

Flournoy, T. (1902) Nouvelles observations sur un cas de somnambulisme avec glossolalia, *Archives de Psychologie*, 1, p. 127.

Forster, J. (1903) *Life of Dickens: Abridged and Revised 3,* London: Chapman & Hall.

Freedman, R. (1996) *Life of a Poet: Rainer Maria Rilke,* New York: Farrar, Straus and Giroux.

Gaillard, P. (1981) *Les Contemplations (1856) analyse critique,* Paris: Hatier.

Garner, S.N., Kahanem, C. & Sprengnether, M. (eds.) (1985) *The (M)other Tongue: Essays in Feminist Psychoanalytic Interpretation,* Ithaca, NY, and London: Cornell UP.

Gass, W.H. (1999) *Reading Rilke: Reflections on the Problems of Translation,* New York: Basic Books.

Guadon, J. (1969) *Le temps de la contemplation: l'oeuvre poétique de Victor Hugo des 'Misères' au 'Seuil de gouffre'* (1845–1856), Paris: Flammarion.

Gaudon, J. & Gaudon, S. (1968) *Présentation. Victor Hugo Oeuvres Complètes. Tome Neuvième. Procès-verbaux des tables parlantes,* Paris: Club français du livre.

Gazzaniga, M.S. (1998) *The Mind's Past,* Berkeley, CA: University of California Press.

Geschwind, D.H., Miller, B.L., DeCarli, C. & Carmelli, D. (2002) Heritability of lobar brain volumes in twins supports genetic models of cerebral laterality and handedness, *PNAS,* 99 (5), pp. 3176–3181.

Gilman, P. (2011) *The Anti-Romantic Child: A Story of Unexpected Joy,* New York: Harper Collins.

Gilman, P. (2013) Early Reader, *New York Times,* 25 August, [Online], http://www.nytimes.com/2013/08/25/books/review/early-reader. html, [accessed 25 August 2013].

Gold, S.N. (2000) *Not Trauma Alone: Therapy for Child Abuse Survivors in Family and Social Context,* Philadelphia, PA: Brunner/Routledge.

Goldberger, Z.D. (2001) Music of the left hemisphere: Exploring the neurobiology of absolute pitch, *Yale Journal of Biology and Medicine,* (74), pp. 323–327.

Golding, W. (1986/1987) *The Man and His Books: A Tribute on His 75th Birthday,* Carey, J. (ed.), New York: Farrar, Straus & Giroux.

Goleman, D. (1990) Brain structure differences linked to schizophrenia in study of twins, *New York Times,* 22 March.

Gorana, P. Machsal, N., Faust, M. & Lavidor, M. (2008) The role of the right cerebral hemisphere in processing novel metaphoric expressions: A transcranial magnetic stimulation study, *Journal of Cognitive Neuro-science*, 20, pp. 170–181.

Grandin, T. (2014) Standing in his shoes. Review: The Reason I Jump: The Inner Voice of a Thirteen-Year-Old Boy with Autism, *The Dana Foundation*, Tuesday 14 February.

Grandin, T. (1995/1996) *Thinking in Pictures*, Foreword Sacks, O., New York: Vintage Books.

Graves, R. (1955/1959) *The Crowning Privilege*, Harmondsworth: Penguin Books.

Graves, R. (1961/1962) *Oxford Addresses on Poetry*, Garden City, NY: Doubleday & Company, Inc.

Graves, R. (1948/1966) *The White Goddess: A Historical Grammar of Poetic Myth*, First American, amend. and enl. ed., New York: Farrar, Straus & Giroux.

Grillet, C. (1929) *Victor Hugo Spirite*, Lyon and Paris: Librairie Emmanuel Vitte.

Grof, S., & Bennett, H.Z. (1993) *The Holotropic Mind: The Three Levels of Human Consciousness and How They Shape Our Lives*, San Francisco, CA: HarperCollins.

Guirdham, A. (1970) *The Cathars and Reincarnation*, London: Spearman.

Hall, J. (2008) *The Sinister Side: How Left-Right Symbolism Shaped Western Art*, Oxford: Oxford UP.

Hacking, I. (1995) *Rewriting the Soul: Multiple Personality and the Sciences of Memory*, Princeton, NJ: Princeton UP.

Hamilton, T. (2009) *Immortal Longings: FWH Myers and the Victorian Search for Life After Death*, Exeter: Imprint Academic.

Hanegraaff, W.J. (1998) *New Age Religion and Western Culture: Esotericism in the Mirror of Secular Thought*, New York: State University of New York Press.

Hanegraaff, W.J. (2001) A woman alone: the beatification of Friederike Hauffe, née Wanner (1801–1829), in *Women and Miracle Stories: A Multi-disciplinary Exploration*, Leiden, Boston, MA, Koln: Brill.

Harper, M.M. (2006) *Wisdom of Two: The Spiritual and Literary Collaboration of George and W.B. Yeats*, Oxford: Oxford UP.

He, W., Chai, H., Zhang, Y., Yu, S., Chen, W. & Wang, W. (2010) Line bisection performance in patients with generalized anxiety disorder and treatment-resistant depression, *Int J Med Sci*, 7 (4), pp. 224–231, [Online], http://www.medsci.org/v07p0224.htm.

Heber, A.S., Fleisher, W.P., Ross, C.A. & Stanwick, R.S. (1989) Dissociation in alternative healers and traditional therapists: A comparative study, *American Journal of Psychotherapy*, 43 (4), pp. 562–574.

Heidegger, M. (1971) *Poetry, Language, Thought*, Hofstadter, A. (trans. & intro.), New York: Harper & Row.

Heller, W., Engels, A.S., Mohanty, A., Herrington, J.D., Nanich, M.T. & Miller, G. (2008) Specificity of regional brain activity in anxiety types during emotion processing, *Brain and Cognition*, 67, S8–S10.

Helmstaedter, C., Kurthen, M., Linke, D.B. & Elger, C.E. (1994) Right hemisphere restitution of language and memory functions in right hemisphere language-dominant patients with left temporal lobe epilepsy, *Brain*, 117 (4), pp. 729–737.

Higgins, E.S. (2008) The new genetics of mental illness, *Scientific American Mind*, (June/July), pp. 41–45.

Hirsch, E. (2002) *The Demon and the Angel: Searching for the Source of Artistic Inspiration*, New York, San Diego, CA, and London: Harcourt, Inc.

Hitt, J. (1999) This is your brain on God, *Wired*, 7 November, pp. 1–5, [Online, accessed 25 August 2014].

Hollis, J. (2000) *The Archetypal Imagination*, College Station, TX: Texas A&M.

Horgan, J. (2003) *Rational Mysticism: Dispatches from the Border Between Science and Spirituality*, Boston, MA, and New York: Houghton Mifflin.

Howell, E.F. (2005) *The Dissociative Mind*, New York and London: Routledge, Taylor and Francis Group.

Hughes, T. (1992) *Shakespeare and the Goddess of Complete Being*, London: Faber and Faber.

Hughes, T. (1995) The Art of Poetry. With Drue Heinz, *The Paris Review*, 71, Spring, [Online], http://www.theparisreview.org/interviews/1669/the-art-of-poetry-no-71-ted-hughes.

Hughes, T. (ed. And intro.) (1996) *A Choice of Coleridge's Verse*, London: Faber and Faber.

Hughes, T. (1998) *Birthday Letters*, New York: Farrar, Straus and Giroux.

Hughes, T. (1999) *Wild Steelhead & Salmon. Winter. Defunct*, [Online], http://www.earth-moon.org/th_intv_steelhead.html

Hughes, T. (2003) *Collected Poems*, Keegan, P. (ed.), New York: Farrar, Straus and Giroux.

Hughes, T. (2007) *Letters of Ted Hughes*, Reid, C. (ed.), New York: Farrar, Straus and Giroux.

Hugo, V. (1856/1972) *Les Contemplations*, Paris: Librairie Générale Française.

Hugo, V. (1864/2003) *William Shakespeare*, Paris: GF Flammarion.

Humphrey, N. (1999) Cave art, autism, and the evolution of the human mind, *Journal of Consciousness Studies*, 6 (6–7), pp. 116–143.

Jabr, F. (2014) Speak for yourself, *Scientific American Mind*, (Jan–Feb), pp. 45–50.

Jackson, D. (1979) DJ: A Conversation with David Jackson. With J.D. McClatchy, *Shenandoah*, 30 (4), pp. 25–44.

James, W. (1994) *The Varieties of Religious Experience: A Study in Human Nature*, New York: The Modern Library.

James, T. (1996) *Dream, Creativity and Madness in Nineteenth-Century France,* Oxford: Clarendon Press.

Jamison, K.R. (1993) *Touched with Fire: Manic-Depressive Illness and the Artistic Temperament,* New York, London, Toronto, Sydney, Tokyo and Singapore: The Free Press.

Janet, P. (1889) *L'automatisme psychologique: Essai de psychologie expérimentale sur les formes inférieures de l'activité humaine,* Paris: Félix Alcan.

Jansen, A., Lohmann H., Scharfe, S., Sehlmeyer, C., Deppe, M. & Knecht, S. (2007) The association between scalp hair-whorl direction, handedness and hemispheric language dominance: Is there a common genetic basis of lateralization?, *Neuroimage,* 35 (2), pp. 853–861.

Jawer, M.A., with Micozzi, M.S. (2009) *The Spiritual Anatomy of Emotion: How Feelings Link the Brain, the Body, and the Sixth Sense,* Rochester, VT: Park Street Press.

Jaynes, J. (1976/1990) *The Origin of Consciousness in the Breakdown of the Bicameral Mind,* Boston, MA: Houghton Mifflin.

Jordison, S. (2013) Sylvia Plath 'didn't want her mother to know she wrote The Bell Jar', *The Guardian,* 19 January, [Online].

Jourdain, R. (1997) *Music, the Brain, and Ecstasy: How Music Captures Our Imagination,* New York: Harper Collins.

Jung, C.G. (1902/1977) *Psychology and the Occult,* Hull, R.F.C. (trans.), Princeton, NJ: Princeton UP.

Jung, C.G. (1961/1989) *Memories, Dreams, Reflections,* Jaffé, A. (ed.), Winston, R. & C. (trans.), New York: Vintage Books.

Jung, C.G. (1997) *Jung on Synchronicity and the Paranormal,* Selected and Intro. R. Main, Princeton, NJ: Princeton UP.

Kane, J. (2004) Poetry as right-hemispheric language, *Journal of Consciousness Studies,* 11 (5–6), pp. 21–59.

Kaplan, R.M. (2006) The neuropsychology of shamanism, *Before Farming,* Article 13, pp. 1–14, [Online], https://www.academia.edu/3549088/The_Neuropsychiatry_of_Shamanism [accessed 18 November 2014].

Keats, J. (1958) *The Letters of John Keats: 1814–1821,* 2 vols, Rollins, H.E. (ed.), Cambridge, MA: Harvard UP.

Keats, J. (1965) *The Keats Circle: Letters and Papers and More Letters and Poems of the Keats Circle,* 2nd ed., Rollins, H.E. (ed.), Cambridge, MA: Harvard UP.

Keats, J. (1990) *John Keats: A Critical Edition of the Major Works,* Cook, E. (ed.), Oxford and New York: Oxford UP.

Kerner, J. (1845) *The Seeress of Prevorst,* Mrs Crowe (trans.), London: scanned courtesy www.spiritwritings.com.

Klein, J. (2008) *Ted Talk,* [Online], http://www.ted.com/talks/joshua_klein_on_the_intelligence_of_crows.html [accessed 19 August 2013].

Knecht, S., Dräger, B., Flöel, A., Lohmann, H., Breitensein, C., Deppe, M., Henningsen, H. & Ringelstein, E.-B. (2001) Behavioural relevance of atypical language lateralization in healthy subjects, *Brain*, 124, pp. 1657–1665.

Koestler, A. (1964/1989) *The Act of Creation,* London, New York, Victoria, Toronto and Auckland: Penguin Books.

Koren, Y. & Negev, E. (2006) *Lover of Unreason: Assia Wevill, Sylvia Platt's Rival and Ted Hughes' Doomed Love,* Cambridge, MA: Da Capo Press.

Kroll, J. (1976/2007) *Chapters in a Mythology: The Poetry of Sylvia Plath,* Gloucestershire: Sutton Publishing Ltd.

Kuipers, E., Garetry, P., Fowler, D., Freeman, D., Dunn, G. & Bebbington, P. (2006) Cognitive, emotional, and social processes in psychosis: refining cognitive behavioural therapy for persistent positive symptoms, *Schizophrenia Bulletin*, 32 (Supplement 1), pp. S24–S31.

Lakoff, G. & Johnson, M. (1999) *Philosophy in the Flesh: The Embodied Mind and its Challenge to Western Thought,* New York: Basic Books.

Lanius, R.A., Williamson, P.C., Bluhm, R.L., Densmore, M., Boksman, K., & Neufeld, R.W.J. (2005) Functional connectivity of dissociative responses in posttraumatic stress disorder: A functional magnetic resonance imaging investigation, *Biological Psychiatry*, 57, pp. 873–884.

Leonard, R. (1999) *The Transcendental Philosophy of Franklin Merrell-Wolff,* Albany, NY: State University Press of New York.

Leonhard, M.A. & Brugger, P. (1998) Creative, paranormal, and delusional thought: A consequence of right hemispheric semantic activation?, *Cognitive and Behavioural Neurology*, 11 (4), pp. 177–183.

LeVay, S. (2011) *Gay, Straight, and the Reason Why: The Science of Sexual Orientation,* New York: Oxford UP.

Lewis, T. (1974) *The Lives of a Cell: Notes of a Biology Watcher,* New York: Penguin Books.

Lindell, A.K. & Jarrad, A.G. (2007) Priming vs. rhyming: Orthographic and phonological representations in the left and right hemispheres, *Annals of Neurology*, April.

Lindell, A.K. (2014) On the interrelation between reduced lateralization, schizotypy, and creativity, *Front. Psychol.*, 5, p. 813, [Online, accessed 26 October 2014].

Longenbach, J. (1988) *Stone Cottage: Pound, Yeats & Modernism,* New York and Oxford: Oxford UP.

Lukoff, D. (1990–91) Divine madness, *Shaman's Drum*, 22, pp. 24–29, [Online], http://www.spiritualcompetency.com/pdfs/sic.pdf [accessed 27 August 2014].

Lurie, A. (2001) *Familiar Spirits: A Memoir of James Merrill and David Jackson,* New York: Penguin Books.

Lyons-Ruth, K. (2003) Disorganized attachment and the relational context of dissociation, *Presented at the 19th Annual Meeting of the International Society for Traumatic Stress Studies*, Chicago, IL, 1 November.

MacLean, K. (2012) Psilocybin and personality change—What do increases in openness tell us about potential mechanisms of action and therapeutic applications?, *Toward a Science of Consciousness*, Tucson, AZ, 14 April.

Maddox, B. (1999) *Yeats's Ghosts: The Secret Life of W.B. Yeats*, New York: HarperCollins.

Maraldi, E. & Krippner, S. (2013) A biopsychosocial approach to creative dissociation: Remarks on a case of mediumistic painting, *Neuroquantology*, 11 (4), pp. 544–572.

Maroda, K. (2004) Sylvia and Ruth, *Salon*, 29 November, [Online], http://www.salon.com/2004/11/29/plath_therapist/ [accessed 21 March 2013].

Martin, L.C. (1923) The poetic mind: by Frederick Clarke Prescott, *International Journal of Psycho-Analysis*, 4, pp. 233–234.

Martz, L.L. (1998) *Many Gods and Many Voices: The Role of the Prophet in English and American Modernism*, Columbia, MI, and London: University of Missouri Press.

Maslow, A.H. (1964) *Religions, Values and Peak-Experiences*, Columbus, OH: Ohio State UP.

Materer, T. (2000) *James Merrill's Apocalypse*, Ithaca, NY, and London: Cornell UP.

McClelland, M.S. & Slaughter, R. (2007) James Merrill: The Doodler, *The Paris Review*, 183, pp. 71–81.

McGilchrist, I. (2009) *The Master and his Emissary: The Divided Brain and the Making of the Western World*, New Haven, CT, and London: Yale UP.

McGuire, W. (ed.) (1974) *The Freud/Jung Letters*, Princeton, NJ: Princeton UP.

McManus, C. (2002) *Right Hand, Left Hand: The Origins of Asymmetry in Brains, Bodies, Atoms and Cultures*, Cambridge, MA: Harvard UP.

Mendez, M.F. (2005) Hypergraphia for poetry in an epileptic patient, *Journal of Neuropsychiatry and Clinical Neuroscience*, 17 (4), pp. 560–561.

Merrill, J. (1979) James Merrill's myth: An interview with Helen Vendler, *The New York Review of Books*, 3 May, pp. 12–13.

Merrill, J. (1982a) The Art of Poetry No. 31. An Interview with J.D. McClatchy, *The Paris Review*, Summer, [Online, accessed 1 February 2011].

Merrill, J. (1982b) James Merrill at home: An interview with Ross Labrie, *Arizona Quarterly*, 38 (1), pp. 19–36.

Merrill, J. (1982/1996) *The Changing Light at Sandover*, New York: Knopf.

Merrill, J. (1986) *Recitative,* McClatchy, J.D. (ed.), San Francisco, CA: North Point Press.

Merrill, J. (1993/1994) *A Different Person: A Memoir,* San Francisco, CA: Harper Collins.

Merrill, J. (2001) *Collected Poems,* McClatchy, J.D. & Yenser, S. (eds.), New York: Knopf.

Middlebrook, D. (2003) *Her Husband: Hughes and Plath – A Marriage,* New York: Viking.

Mikal, E.J. (2005) *Until Darkness Holds No Fear: Healing a Multiple Personality,* Boulder, CO: Books Beyond Borders, Inc.

Mitchell, J. (2000) *Mad Men and Medusas: Reclaiming Hysteria,* New York: Basic Books.

Moffett, J. (1984) *James Merrill: An Introduction to the Poetry,* New York: Columbia UP.

Molnar-Szakacs, I., Iacoboni, M., Koski, L., *et al.* (2005) Functional segregation within pars opercularis of the inferior frontal gyrus: Evidence from fMRI studies of imitation and action observation, *Cerebral Cortex,* 15 (7), pp. 986–994.

Morrison, B. (2001) Keeper of a stubborn faith, *The Guardian,* 27 October.

Mühl, A.M. (1929) Automatic writing as an indicator of the fundamental factors underlying the personality, *An Outline of Abnormal Psychology,* St. Elizabeth's Hospital, Washington, DC, pp. 162–183.

Murphy, M. (1992) *The Future of the Body: Explorations into the Further Evolution of Human Nature,* Los Angeles, CA: Tarcher.

Mutigny, J. de. (1981) *Victor Hugo et le spiritisme,* Paris: Editions Fernand Nathan.

Myers, F.W.H. (1903) *Human Personality and Its Survival of Bodily Death,* New York: Longmans, Green and Co.

Narada, T. (1992) *A Manual of Buddhism,* Malaysia: Buddhist Missionary Society, [Online], http://www.khandro.net/Bud_mother.htm.

Neumann, E. (1959/1974) *Art and the Creative Unconscious,* Princeton, NJ: Princeton UP, Bollingen Series LXI.

Newberg, A.& Waldman, M.R. (2006/2007) *Born to Believe: God, Science, and the Origin of Ordinary and Extraordinary Beliefs,* New York, London, Toronto and Sydney: Free Press.

Nauert, R. (2010) *Brain Region Linked to Introspective Thought,* [Online], http://psychcentral.com/news/2010/09/20/brain-region-linked-to-introspective-thoughts/18434.html [accessed 20 September 2010].

Ornstein, R. (1972/1996) *The Psychology of Consciousness,* New York: Arkana.

Ornstein, R. (1997) *The Right Mind: Making Sense of the Hemispheres,* New York, San Diego, CA, and London: Harcourt Brace & Company.

Orr, G. (2002/2004) *The Blessing: A Memoir*, San Francisco, CA, and Tulsa, OK: Council Oak Books.

Ostmeier, D. (2000) Gender debates between Rainer Maria Rilke and Lou Andreas-Salomé, *The German Quarterly*, 73 (3), pp. 237–252.

Oubré, A.Y. (1997) *Instinct and Revelation: Reflections on the Origins of Numinous Perception*, vol. 10, Amsterdam: Gordon and Breach Publishers.

Owen, I.M. (1976) *Conjuring up Philip: An Adventure in Psychokinesis*, New York: Harper & Row.

Panegyres, M. (2012) The concept of thumos and its role in Ancient Greek and contemporary modes of resistance, [Online, accessed 28 December 2013].

Peres, J.F., *et al.* (2012) Neuroimaging during trance state: A contribution to the study of dissociation, *PLOS ONE*, 7 (11), 16 Nov, [Online], no pag.

Persinger, M. (1987) *Neuropsychobiological Bases of God Beliefs*, New York, Westport, CT, and London: Praeger.

Persinger, M.A. & Makarec, K. (1987) Temporal lobe epileptic signs and correlative behaviors displayed by normal populations, *J Gen Psychol.*, 114 (2), pp. 179–195.

Persinger, M.A. & Makarec, K. (1992) The feeling of a presence and verbal meaningfulness in context of temporal lobe function: Factor analytic verification of the muses?, *Brain and Cognition*, 20, pp. 217–226.

Persinger, M.A. & Healey, F. (2002) Experimental facilitation of the sensed presence: Possible intercalation between the hemispheres induced by complex magnetic fields, *Journal of Nervous & Mental Disease*, 190 (8), pp. 533–541.

Picard, F. & Craig, A.D. (2009) Ecstatic epileptic seizures: A potential window on the neural human basis of self-awareness, *Epilepsy & Behavior*, 16, pp. 539–546.

Pinet, H. (1992) *Rodin: Hands of Genius*, Palmer, C. (trans.), New York: Abrams.

Pinker, S. (1997) *How the Mind Works*, New York and London: W.W. Norton & Company.

Plath, S. (1960/2008) *The Collected Poems: Sylvia Plath*, Hughes, T. (ed. and intro.), New York: Harper Perennial Modern Classics.

Plath, S. (1971) *The Bell Jar*, New York: Bantam.

Plath, S. (1975) *Letters Home: Correspondence 1950-1963*, Plath, A.S. (sel. and ed.), New York: Harper Perennial.

Plath, S. (2000) *The Unabridged Journals of Sylvia Plath 1950–1962*, Kukil, K.V. (ed.), New York: Anchor Books.

Plath, S. (2010) *The Spoken Word: Sylvia Plath*, British Library.

Platt, C.B. (2007) Presence, poetry and the collaborative right hemisphere, *Journal of Consciousness Studies*, 14 (3), pp. 36–53.

Poincaré, H. (1913) *Mathematical Creation: The Foundations of Science,* Halsted, G.B. (trans.), New York: The Science Press.

Polito, R. (1994) *A Reader's Guide to James Merrill's The Changing Light at Sandover,* Ann Arbor, MI: University of Michigan Press.

Prescott, F.C. (1922) *The Poetic Mind,* Westport, CT: Greenwood Press.

Previc, F.H. (2005) The role of the extrapersonal brain systems in religious activity, *Consciousness and Cognition,* [Online, accessed 9 September 2009].

Previc, F.H. (2009) *The Dopaminergic Mind in Human Evolution and History,* Cambridge: Cambridge UP.

Prichard, E., Propper, R.E. & Christman, S.D. (2013) Degree of handedness, but not direction, is a systematic predictor of cognitive performance, *Front. Psychol.,* 4 (9), no pag.

Propper, C., Moore, G.A., Mills-Koonce, R., Halpern, C.T., Hill-Soderlund, A.L., Calkins, S.D., Carbone, M.A. & Cox, M. (2008) Gene-environment contributions to the development of infant vagal reactivity: The interaction of dopamine and maternal sensitivity, *Child Development,* 79 (5), pp. 1377–1394.

Putnam, F.W. (1992) Discussion: Are alter personalities fragments or figments?, *Psychoanalytic Inquiry,* 12, pp. 95–111.

Putnam, F.W. (1997) *Dissociation in Children and Adolescents,* New York: Guilford Press.

Radin, D. (2006) *Entangled Minds: Extrasensory Experiences in a Quantum Reality,* New York: Paraview Pocket Books.

Ramachandran, V.S. & Blakeslee, S. (1998) *Phantoms in the Brain: Probing the Mysteries of the Human Mind,* New York: Quill William Morrow.

Rawlings, D. & Locarnini, A. (2006) Dimensional schizotypy, autism and unusual word associations in artists and scientists, *J Res Pers,* 42 (2), pp. 465–471.

Read, J., Os, J., Morrison, A.P. & Ross, C.A. (2005) Childhood trauma, psychosis and schizophrenia: A literature review with theoretical and clinical implications, *Acta Psychiatrica Scandinavica,* 112, pp. 330–350.

Reinders, A.A.T.S., Nijenhuis, E.R.S., Paans, A.M.J., Korf, J., Willemsem, A.T.M. & den Boer, J.A. (2003) One brain, two selves, *NeuroImage,* 20 (4), pp. 2119–2125.

Reinders, A.A.T.S., Nijenhuis, E.R., Quak, J., Korf, J., Haaksma, J., Paans, A.M.J., Willemsen, A.T.M & den Boer, J.A. (2006) Psychobiological characteristics of dissociative identity disorder: A symptom provocation study, *Biological Psychiatry,* 60, pp. 730–740.

Rilke, R.M. (1949/1964) *The Notebooks of Malte Laurids Brigge,* Herter, M.D. (trans.), New York and London: W.W. Norton & Company.

Rilke, R.M. (1987) *The Sonnets to Orpheus, by Rainer Maria Rilke,* Mitchell, S. (trans.), New York: Simon and Schuster.

Rilke, R.M. (2009) *The Poetry of Rilke*, Snow, E. (trans. and ed.), New York: North Point Press.

Robb, G. (1997/1998) *Victor Hugo: A Biography*, London and New York: W.W. Norton & Co.

Rogers, A.G. (1995) *A Shining Affliction: A Story of Harm and Healing in Psychotherapy*, New York: Penguin Books.

Rogers, A.G. (2007) *The Unsayable: The Hidden Language of Trauma*, New York: Ballantine Books.

Rohls, P. & Ramírez, J.M. (2006) Aggression and brain asymmetries: A theoretical review, *Aggression and Violent Behavior*, 11 (3), pp. 283–297.

Roman, S. & Packer, D. (1984) *Opening to Channel: How to Connect with Your Guide*, Tiburon, CA: H.J. Kramer, Inc.

Rosenberg, D. (2000/1) Speaking Martian, *Cabinet Magazine Online (Invented Languages)*, (1), [Online, accessed 5 May 2008].

Ross, C.A. (1989) *Multiple Personality Disorder: Diagnosis, Clinical Features, and Treatment*, New York: John Wiley & Sons.

Ross, C.A. (1994) *The Osiris Complex: Case Studies in Multiple Personality Disorder*, Toronto: University of Toronto Press.

Sabbah, P., Chassoux, F., Leveque, C., Landre, E., Baudoin-Chial, S., Devaux, B., Mann, M., Godon-Hardy, S., Nioche, C., Aït-Ameur, A., Sarrazin, J.L., Chodkiewicz, J.P. & Cordoliani., Y.S. (2000) Functional MR imaging in assessment of language dominance in epileptic patients, *Epilepsia*, 41 (Suppl 7), p. 62.

Saddlemyer, A. (2002) *Becoming George: The Life of Mrs W.B. Yeats*, Oxford: Oxford UP.

Sacks, O. (2012) *Hallucinations*, New York and Toronto: Alfred A. Knopf.

Sagar, K. (2010/2012) *The Laughter of Foxes: A Study of Ted Hughes*, Liverpool: Liverpool UP. Website: http://keithsagar.co.uk/tedhughes.html

Sagar, K. (2011) Ted Hughes and the divided brain, presentation at the *International Conference on Ted Hughes*, Pembroke College, Cambridge, September 2010. First published in the *Journal of the Ted Hughes Society*, 1, Summer.

Sanders, N.K. (1960/1972) *The Epic of Gilgamesh*, Harmondsworth: Penguin Books.

Sarde, M. (1983) *Regard sur les Françaises*, Paris: Stock.

Savarese, R. J. (2008) The lobes of autobiography: Poetry and autism, *Stone Canoe*, 2 (Spring), [Online, accessed 1 August 2014].

Savic, I. & Lindstrom, P. (2008) PET and MRI show differences in cerebral asymmetry and functional connectivity between homo- and heterosexual subjects, *PNAS*, 16 June, doi: 10.1073/pnas.080156.

Savitz, J., Solms, M., Pietersen, E., Ramesar, R. & Flor-Henry, P. (2004) Dissociative Identity Disorder associated with mania and change in handedness, *Cognitive & Behavioral Neurology*, 17 (4), pp. 233–237.

Schaafsma, P. (1980) Trance and transformation in the canyons: Shamanism and early rock art on the Colorado plateau, in *Indian Rock Art of the Southwest*, Santa Fe School of American Research.

Schore, A.N. (1994) *Affect Regulation and the Origin of the Self: The Neurobiology of Emotional Development*, Hillsdale, NJ: Lawrence Erlbaum Associates.

Schore, A.N. (2009) Relational trauma and the developing right brain: An interface of psychoanalytic self psychology and neuroscience, *Self and Systems: Annals of The New York Academy of Sciences*, 1159, pp. 189–203.

Schore, A.N. (2010) Relational trauma and the developing right brain: The neurobiology of broken attachment bonds, in Baradon, T. (ed.) *Relational Trauma in Infancy: Psychoanalytic Attachment and Neuropsychological Contributions to Parent–Infant Psychotherapy*, pp. 19–47, London: Routledge.

Schneps, M.H. (2014) The advantages of dyslexia: With reading difficulties can come other cognitive strengths, *Scientific American*, 19 August, [Online], http://www.scientificamerican.com/article/the-advantages-of-dyslexia/ [accessed 19 August 2014].

Schreber, D.P. (1955/2000) *Memoirs of My Nervous Illness*, New York: Review Books.

Schuchard, M.K. (2006/2008) *William Blake's Sexual Path to Spiritual Wisdom*, Rochester, NY: Inner Traditions.

Segal, S. (1996) *Collision with the Infinite*, San Diego, CA: Blue Dove Press.

Selfe, L. (2011) *Nadia Revisited: A Longitudinal Study of an Autistic Savant*, New York: Psychology Press.

Seo, D., Patrick, C.J. & Kennealy, P.J. (2008) Role of serotonin and dopamine system interactions in the neurobiology of impulsive aggression and its comorbidity with other clinical disorders, *Aggressive Violent Behavior*, 5, pp. 383–395.

Sewell, R. (1960/1971) *The Orphic Voice: Poetry and Natural History*, New York: Harper Torch Books.

Seymour-Smith, M. (1982/1988) *Robert Graves: His Life and Works*, New York: Paragon House Publishers.

Shamdasani, S. (2011) Jung after the Red Book, Lecture, *Jung Center*, Houston, TX, 2 April.

Shanon, B. (2002) *The Antipodes of the Mind: Charting the Phenomenology of the Ayahuasca Experience*, Oxford: Oxford UP.

Shaywitz, S.E. & Shaywitz, B.A. (2005) Dyslexia (specific reading disability), *Biological Psychiatry*, pp. 1301–1309.

Sheehan, D. & Reffell, P. (2013) *BrainLines: The 7 Personality Types that Determine How We Love, Live and Work*, Arts, J. (fwd.), CreateSpace.

Shenton, M.E., Kikinis, R., Jolesz, F.A., Pollak, S.D., LeMay, M., Wible, C.G., Hokama, H., Martin, J., Metcalf, D., Coleman, M. & McCarley, R.W. (1992) Left temporal lobe abnormalities in schizophrenia and thought disorder: A quantitative MRI study, *New England Journal of Medicine*, 327, pp. 604–612.

Shlain, L. (1998) *The Alphabet Versus the Goddess: The Conflict Between Word and Image*, New York: Penguin Arkana.

Shlain, L. (2014) *Leonardo's Brain: Understanding Da Vinci's Creative Genius*, Guilford, CT, and Helena, MT: Lyons Press.

Shorto, R. (1999) *Saints and Madmen: Psychiatry Opens Its Doors To Religion*, New York: Henry Holt and Company.

Siegel, D. (1999) *The Developing Mind: Toward a Neurobiology of Interpersonal Experience*, New York: Guilford Press.

Simon, G. (1923/1996) *Chez Victor Hugo: Les Tables tournantes de Jersey*, Paris: L'école des lettres.

Skea, B.R. (2006) *A Jungian Perspective on the Dissociability of the Self*, [Online], www.cgjungpage.org. Presented 3 February 1995 at the C.G. Jung Education Center, Pittsburgh, PA.

Smeets, G., Merchelbach, H. & Griez, E. (1997) Panic disorder and right-hemisphere reliance, *Anxiety, Stress & Coping: An International Journal*, 10 (3), pp. 245–255, [Online], 29 May 2007, http://www.tandfonline.com/doi/abs/10.1080/10615809708249303.

Somers, M., Sommer, V., Boks, M.P. & Kahn, R.S. (2009) Hand-preference and population schizotypy: A meta analysis, *Schizophrenia Research*, 108 (1–3), pp. 25–32.

Sommer, I.E.C. & Kahn, R.S. (eds.) (2009) *Language Lateralization and Psychosis*, Cambridge: Cambridge UP.

Sperry, S.M. (1973/1994) *Keats the Poet*, Princeton, NJ: Princeton UP.

Springer, S.P. & Deutsch, G. (1981/1998) *Left Brain, Right Brain: Perspectives from Cognitive Neuroscience*, 5th ed., New York and Basingstoke: W.W. Freeman and Company.

Sritharan, A., Line, P., Sergejew, A., Silberstein, R., Egan, G. & Copolov, D. (2005) EEG coherence measures during auditory hallucinations, *Psychiatry Research*, 136 (2–3), pp. 189–200. Cited in Kuijsten, M. (ed.) (2006) *Reflections on the Dawn of Consciousness: Julian Jaynes's Bicameral Mind Theory Revisited*, Henderson, NV: Julian Jaynes Society.

Stevenson, R.L. (1892) A chapter on dreams, *Across the Plains, with Other Memories and Essays*, New York: Scribner's Sons.

Suomi, S.J. (2005) Aggression and social behaviour in rhesus monkeys, in Bock, G. & Goode, J. (eds.) *Molecular Mechanisms Influencing Aggressive Behaviours*, Novartis Foundation, 268, pp. 216–226.

Swedenborg, E. (1888) *The Swedenborg Concordance 1,* Faulkner Potts, Rev. J. (ed. and trans.), Edinburgh: Morrison and Gibb.

Sword, H. (1995) *Engendering Inspiration: Visionary Strategies in Rilke, Lawrence, and H.D.,* Ann Arbor, MI: University of Michigan Press.

Sword, H. (2002) *Ghostwriting Modernism,* Ithaca, NY, and London: Cornell UP.

Tang, G., Gudsnuk, K., Sheng-Han, K., Cotrina, M.L., Gorazd, R., Sosunov, A., Sonders, M.S., Kanter, E., Castagna, C., Yamamoto, A., Yue, A., Ottavio, P., Bradley S., Champagne, F., Dwork, A.J., James, G. & Sulzer, D. (2014) Loss of mTOR-dependent macroautophagy causes autistic-like synaptic pruning deficits, *Neuron,* http://dx.doi.org/10.1016/j.neuron. 2014.07.040 [accessed 23 August 2014].

Taylor, J. (1998) *The Living Labyrinth: Exploring Universal Themes in Myths, Dreams, and the Symbolism of Waking Life,* New York and Mahwah, NJ: Paulist Press.

Taylor, K.I. & Regard, M. (2003) Language in the right cerebral hemisphere: Contributions from reading studies, *News in Physiological Sciences,* 18, pp. 257–261.

Trask, W. (1996) *Joan of Arc in Her Own Words,* New York: Books & Co./A Turtle Point Imprint.

Trimble, M.R. (2007) *The Soul in the Brain: The Cerebral Basis of Language, Art, and Belief,* Baltimore, MD: The Johns Hopkins UP.

Thomas, L. (1974/1978) *The Lives of a Cell,* New York: Viking.

Tweedy, R. (2013) *The God of the Left Hemisphere: Blake, Bolte Taylor and the Myth of Creation,* London: Karnac.

Vacquerie, A. (1863) *Les miettes de l'histoire,* Paris: Pagnerre.

Van Lommel, P. (2010) *Consciousness Beyond Life: The Science of the Near-Death Experience,* New York: HarperOne.

Venkatasubramanian, G., Jayakumar, P.N, Hongasandra, R.N., Nagaraja, D., Deeptha, R. & Ganadadhar, B.N. (2008) Investigating paranormal phenomena: Functional brain imaging of telepathy, *International Journal of Yoga,* 1 (2), pp. 55–71.

Vermetten, E., Schmahl, C., Linder, S., Loewenstein, R.J., & Bremner, J.D. (2006) Hippocampal and amygdalar volumes in dissociative identity, *American Journal of Psychiatry,* 163, pp. 630–636.

Vikingstad, E.M., George, K.P., Johnson, A.F. & Cao, Y. (2000) Cortical language lateralization in right-handed normal subjects using functional magnetic resonance imaging, *Journal of the Neurological Sciences,* 175 (1), pp. 17–27.

Waldvogel, B., Ullrich, A. & Strasburger, H. (2007) Blind und Sehend in Einer Person, *Nervarzt,* 11, pp. 1303–1309.

Walsch, N.D. (1995) *Conversations with God,* New York: G.P. Putnam's Sons.

Wapner, J. (2008) Blogging—It's good for you: The therapeutic value of blogging becomes a focus of study, *Scientific American Mind*, 19 May, [Online, accessed 18 September 2014].

Ware, L.A. (2014) Are face-blindness and synesthesia linked to autism spectrum disorders?, *The Dana Foundation*, 19 March, [Online, accessed 4 January 2014].

Washington, P. (1993) *Madame Blavatsky's Baboon: A History of Mystics, Mediums, and Misfits Who Bought Spiritualism to America*, New York: Schocken Books.

Wegner, D.M. (2003) The mind's best trick: How we experience conscious will, *Trends in Cognitive Sciences*, 7 (2), pp. 6–69.

Weinstein, S. & Graves, R.E. (2001) Creativity, schizotypy and laterality, *Cognitive Neuropsychiatry*, 6 (2), pp. 131–146. Cited in Kaplan (2006).

Weiss, B.L. (1988) *Many Lives, Many Masters*, New York: Simon & Schuster.

Weissman, J. (1993) *Of Two Minds: Poets Who Hear Voices*, Hanover and London: Wesleyan UP.

West, T.G. (1997) *In the Mind's Eye: Visual Thinkers, Gifted People with Dyslexia and Other Learning Difficulties, Computer Images and the Ironies of Creativity*, updated ed., New York: Prometheus Books.

Whitley, D.S. (2009) *Cave Paintings and the Human Spirit: The Origin of Creativity and Belief*, Amherst, NY: Prometheus Books.

Whitley, D.S. & Whitley, C.M. (2015) The origins of artistic genius and the archaeology of emotional difference, in *The Genesis of Creativity and the Origin of the Human Mind*, Putova, B. & Soukup, V. (eds.), Prague: Karolinum House Publishing.

Whitman, W. (2001) *Leaves of Grass. The 'Death-Bed' Edition*, Williams, W.C. (intro.), New York: The Modern Library.

Winkelman, M. (2010) *Shamanism: A Biopsychosocial Paradigm of Consciousness and Healing*, Santa Barbara, CA: Praeger.

Woodman, M. (1990) *The Ravaged Bridegroom: Masculinity in Women*, Toronto: Inner City Books.

Woolacott, I.O.C., Fletcher, P.D., Massey, L.A., Pasupathy, A., Rossor, M.N., Caine, D., Rohrer, J.D. & Warren, J.D. (2014) Compulsive versifying after treatment of transient epileptic amnesia, *Neurocase: The Neural Basis of Cognition*, London: Routledge.

Woolger, R.J. (1988) *Other Lives, Other Selves: A Jungian Psychotherapist Discovers Past Lives*, New York: Bantam Books.

Wulff, D.M. (1997) *Psychology of Religion: Classic and Contemporary*, 2nd ed., New York: John Wiley and Sons.

Yeats, W.B. (1916) *Reveries over Childhood and Youth*, New York: Macmillan Company, reproduction BiblioLife, LLC.

Yeats, W. B. (1922/2010) *The Trembling of the Veil*, Project Gutenberg.

Yeats, W.B. (1925/2008) *The Collected Works of W.B. Yeats. XIII. A Vision,* Paul, C.E. & Mills Harper, M. (eds.), New York, London, Toronto and Sydney: Scribner.

Yeats, W.B. (1937/1973) *A Vision,* New York: Collier Books.

Yeats, W.B. (1966/1971) *Selected Poems and Two Plays of William Butler Yeats,* Rosenthal, M.L. (ed. and intro.), New York: Collier Books.

Yenser, S. (1987) *The Consuming Myth: The Work of James Merrill,* Cambridge, MA: Harvard UP.

Yuan, W., Szaflarski, J.P., Schmithorst, V.J., Schapiro, M., Byars, A.W., Strawsburg, R.H. & Holland, S.K. (2006) FMRI shows atypical language lateralization in pediatric epilepsy patients, *Epilepsia,* 47 (3), pp. 593–600.

CPSIA information can be obtained at www.ICGtesting.com
Printed in the USA
BVOW11s0206100915

417400BV00004B/37/P